TIGER BEETLES OF ALBERTA

Killers on the Clay, Stalkers on the Sand

Tiger Beetles of Alberta

John Acorn

Killers on the Clay, Stalkers on the Sand

THE UNIVERSITY OF ALBERTA PRESS

Published by
The University of Alberta Press
Ring House 2
Edmonton, Alberta T6G 2E1

Printed in Canada 5 4 3 2 1

Printed and bound by Speedfast Colour Press Inc., Edmonton, Alberta.
∞ Printed on acid-free paper.

CANADIAN CATALOGUING IN PUBLICATION DATA

Acorn, John Harrison, 1958–
The tiger beetles of Alberta

(Alberta insects series)
Includes bibliographical references.
ISBN 0–88864–345–4

1. Tiger beetles—Alberta. I. Title. II. Series.
QL596.C56A26 2001 595.76'2 C00-910328-7

The University of Alberta Press gratefully acknowledges the support received for its program from the Canada Council for the Arts. The Press also acknowledges the financial support of the Government of Canada through the Book Publishing Industry Development Program for its publishing activities.

To my wife Dena and our son Jesse, whether he grows up liking beetles or not.

Contents

Preface to the Alberta Insects Series

I FIND INSECTS FASCINATING, and I often despair at how few people share my passion for these ubiquitous, bizarre animals. After puzzling over this thought for 35 years as a "bug guy," I am now pretty sure I understand the underlying problem.

Entomologists are like any other group of specialists in that they have created a jargon. To them, the language they use is an unquestionable part of the study and appreciation of insects. To the newcomer, however, this language is a barrier and a hindrance. So, I have decided to remove the barrier wherever I can. I use words like forehead and wing cover, rather than frons and elytron. After all, they mean precisely the same thing.

At the same time, too many books about insects fail to include pictures of the entire bug, head to toe, in colour, showing what it looks like when it is alive. Many entomologists do not anticipate that most people interested in nature won't be inspired by diagrams of this or that fragment of an insect's body. Books illustrated only with diagrams of genitalia dissections are especially incomprehensible to the non-specialist. Admittedly, some insects are very difficult, if not impossible, to distinguish at a glance, but my impression has been that the average bug-loving person still wants to know what the whole critter looks like. Some may disagree, but I think that encouraging the elitists to open up is just as much a part of promoting insect awareness as helping the newcomers expand their awareness.

So to heck with the traditional approach! I want to write books for nature-loving people of all sorts, and I want the books to be useful from the moment you open the covers. I want to share the beautiful colours and shapes that the insects display, as well as the fun I have had searching for them and observing their lives. I hope you agree.

The book you now hold is intended to be the first volume in a series about Alberta's amazing insects. Future volumes will explore damselflies, dragonflies, ladybugs, and larger moths. Watch for them!

Acknowledgements

MANY PEOPLE DESERVE CREDIT for their part in the development of this book. My parents encouraged my early interest in tiger beetles, as did Kate, Don, and Pat Shaw, Gary Noonan, and George Ball. More recently, John Spence served as a fine thesis supervisor, and John and Bert Carr have been wonderfully generous with their time and expertise. My wife Dena Stockburger has come to share my interest in the tigers, and has helped out many times in the field. Other field companions who contributed significantly to this project include Deena Arnold, Su-Ling Goh, Shelley Humphries, Chris Fisher, Carole Patterson, and Christina Hall. I also benefitted greatly by the contributions of Vanessa Block, Neil Evenhuis, Robert Graves, David Gummer, Ron Huber, Kate Kelley, Mike Kippenhan, David Maddison, Tom Schultz, Felix Sperling, and Stephanie Stephens. The text was reviewed in whole or part by George Ball, Rick Freitag, Gerry Hilchie, Janet Sisson, Chris Fisher, and Carole Patterson, but of course any remaining errors or outlandish opinions are entirely my own doing.

Large or small, all tiger beetles have a distinctive body shape.

Chicken Chokers and Anchor Backs

AN INTRODUCTION TO TIGER BEETLES

AMONG BEETLES, tiger beetles are second only to ladybugs, and perhaps fireflies, as favourites of nature-loving people. They are large, as beetles go (a big one is about 15 mm long, or 0.6 inches), active by day in places where they are easy to see, and many have lovely colours and interesting markings on their wing covers. They are as watchable as birds, and easily as fascinating.

To recognize a tiger beetle, look for large eyes, which make the head wider than the PRONOTUM (the top of the body between the head and the wing covers), long threadlike antennae, and long slender legs. There are a few ground beetles that look a bit like tiger beetles, and anyone who looks for beetles on beaches will eventually come upon *Elaphrus* or *Opisthius*, both of which have heavily textured wing covers and shorter legs than a real tiger beetle. One entomologist friend of mine, when he first saw an *Elaphrus*, thought for sure he had discovered a new sort of tiger beetle, which he planned to name "*Microcicindela punctulata*," meaning "the little tiger beetle with tiny pits." I dare say I once had that same fleeting glimpse of scientific glory myself.

The only other beetles that even vaguely resemble tiger beetles are some of the leaf beetles (in the family Chrysomelidae) and the checkered beetles (in the family Cleridae), but these creatures live on plants and apart from their wide, big-eyed heads they aren't very tiger beetlish at all.

Tiger beetles have their own folklore and a bit of human culture associated with them. For example, "chicken chokers" don't really choke chickens, but you can't blame rural people in the American southeast for thinking that the larvae of tiger beetles might try from time to time. (The larvae have fearsome mandibles, and they lie in wait for much smaller prey, flush with the ground surface in the sorts of open, flat places where chickens like to peck.) Tiger beetles themselves don't seem to have attracted a lot of nicknames—perhaps since "tiger beetle" is such a fine name to begin with. My friend Chris Fisher, however, did start calling them "anchor backs" when he first became a tiger-beetler, in reference to the shape of the dark markings on their wing covers.

Other tiger-beetle folk beliefs are few in western Canada, but even among the entomological crowd people hold an exaggerated opinion regarding the power of a tiger beetle's bite, and their reputation far outstrips their abilities to do harm. Our son Jesse said it well, when he was a mere 20 months old. One of his favourite books was a photographic guide to German ground beetles and tiger beetles (*Laufkäfer: Beobachtung, Lebensweise*), and when I

(top) Almost a tiger beetle, but not quite: the ground beetle Elaphrus.
(bottom) The face of a tiger beetle leaves a lasting impression.

A bleak afternoon at the once-magnificent tiger-beetle haven, Aspen Beach at the summer village of Gull Lake.

showed him a head-on picture of a sylvan tiger beetle (*Cicindela sylvatica*), he called it "scare beetle." No kidding.

The intent of this book is to bring you under the influence of these beetles and their cult of followers, and to allow the phenomenon of cicindelophily (the love or admiration of tiger beetles) to spread to minds other than my own and those of my friends. Tiger beetles will probably never be as important to our society as are bears, wolves, or eagles, but they do have a gut-level appeal. Those of us who know them well love tiger beetles the way birders love birds or orchid-growers love orchids (two fine passions, if I do say so myself).

My own earliest memories of tiger beetles go back to family vacations at Aspen Beach on Gull Lake. In the summer of 1971, while running down the sandy path from my aunt's cottage to the shore, I would stir up hundreds of them. When I took the time to catch a few, I was delighted with their crisp markings, their big jaws and eyes, and their bright, iridescent blue-green undersides—it was, after all, the combination of striped patterns and ferocity that gave tiger beetles their name. I was thirteen years old at the time, just graduating from butterfly collecting to the appreciation of the rest of nature.

The following winter, while skulking in the library (I was a proudly nerdish teenager), I spotted a thin green book on the shelf: *The Cicindelidae of Canada* by J.B. Wallis of Winnipeg. I took it down, opened it, and beheld four plates of colour photos showing every species of tiger beetle in the country. By today's standards, they were mediocre illustrations, but many people I have met since were equally inspired by the legless, antennaless, pinhole-through-the-body tiger beetles in Wallis' book. After all, it was the only generally available text that allowed one to identify any group of beetles to the species level in this part of the world. We revered the book, to us it was the Beetle Bible.

When spring came, I grabbed my net and went out in search of the tigers. Soon, my childhood friends Bob Davidson, Mike Chrustawka, and Kate Shaw accompanied me. We called the beetles "cic" (pronounced "sick" and short for *Cicindela*), and we loved to say "let's go get sick" in front of our parents. Kate's folks saw past the joke and soon came to love tiger beetles themselves. Before long, the Shaw family was planning their summer vacations around classic tiger beetle locations.

In letters from fellow junior entomologists, I learned that tiger-beetle mania was also afflicting another group of teenagers in Calgary, centred around Gerry Hilchie and Ted Pike. Gerry and Ted were in turn mentored by John and Bert Carr, an amateur beetle-collecting couple. John Carr's father, Frederick S. Carr, was also a beetle collector, and his collection is now housed at the University of Alberta, where it is still used frequently by many beetle enthusiasts, including me.

In 1972, I caught some rare ground beetles while trying to dig tigers out of their overnight burrows at Gull Lake, and Dr. George Ball tracked me down to ask where I got the specimens. He was then Chairman of the Entomology Department at the University of Alberta and is still a mentor to me today. Meeting him was an important moment for me, and I was amazed to learn that he was not only responsible for seeing Wallis' book to print (Wallis was in poor health and died shortly after it was published) but also added his masterful touch to the manuscript. With characteristic modesty, he declined full credit for his contributions, and the book remains "Wallis" to tiger-beetlers, not "Wallis and Ball."

In a way, this book is intended as a modernized, localized version of Wallis' volume. In the pages that follow I not only want to update our knowledge of the tiger beetles of Alberta—and tiger-beetle science is a fast-expanding field—but also to celebrate their beauty, their remarkable adaptations, their quirks and personalities, and the grandeur of the landscapes they inhabit. Living in unvegetated, geologically active environments—ground that many people consider waste places—tiger beetles serve as symbols of the value of even the most bare and lifeless spots on the planet. These are the kinds of places that make the west "The West"—places where only a tiger beetle, or someone in search of tiger beetles, would spend time.

About 115 species of tiger beetles inhabit North America north of Mexico, and 19 of these live in Alberta. Saskatchewan and Manitoba have about the same number, but no other province or Canadian territory comes close. Here in Alberta, we also have a long and distinguished legacy of tiger-beetle science and appreciation. This book is one product of that legacy, and of all the good friendships and sunny field-trips that go with it. It is my hope that it will be useful to everyone from beetle-loving kids to working entomologists.

If every reader of this book takes the time to find just one tiger beetle in the field, I will count the project successful. Of course, if you merely read the text and admire the beetles in the photographs, that's all right with me too. I

(top) John and Bert Carr, the honourary grandparents of all Alberta tiger-beetlers, posing in their beetle room. (bottom) Frederick S. Carr, Alberta's tiger-beetling pioneer, and his dog Spot, in 1929.

Vanessa Block, entomology student, wears her love of tiger beetles proudly on her right shoulder.

This is not a lifeless landscape or a place for off-road vehicles—this is a tiger-beetle habitat!

hope you will learn some things about tiger beetles, beetles in general, insects in general, the science of biology, and the province of Alberta. Seeing the world through the eyes of a tiger-beetle specialist may give you a new slant on things, or at least a different one than what most folks are used to.

One more thing—don't make the mistake of thinking that because you are searching for beetles, the outings will be tame. One friend of mine fell down a cliff while searching for tiger beetles and broke her leg. I myself have encountered angry rattlesnakes, fresh grizzly tracks, irate ranchers with high-powered rifles, and on a number of occasions I have also come very close to suffering heat stroke. Another friend of mine, Phyllis Pineda, wrote to me not long ago saying, "during my last trip down to the Great Sand Dunes [in Colorado, not Saskatchewan], my field volunteer and I were tracked by 2 mountain lions for the 4 days we were in the field! I mean, they were right behind us. We would go back over our tracks sometimes within 45 minutes, and sure enough we would see their tracks on top of ours! Yikes! We never did see them, though. The resource specialist at the Great Sand Dunes suggested I title my thesis: 'Hunting Tigers, Tracked by Lions'." If this can happen in Colorado, it can happen here too!

(top) The Bengal of Alberta's beetles: a beautiful tiger beetle from the Empress sands.
(bottom) A bronzed tiger beetle, mercilessly attacking a much larger winter moth.

1
The Life of a Tiger Beetle

IT'S 6 O'CLOCK IN THE MORNING, mid August, and the sun is rising into a cloudless indigo sky. From the back of a shallow burrow, a little more than 10 cm long, a tiger beetle watches and waits. She is a beautiful tiger beetle (*Cicindela formosa*), from the dunes near Empress, Alberta.

The sun warms the sand, the dew evaporates, and at about 8 o'clock the surface temperature reaches 20°C (68°F). Now it is time to get up. Out comes the beetle, keeping low to the ground for warmth. Nearby, others emerge from their own sandy burrows. The ones whose tunnels have collapsed on them will dig out a bit later, when they are warmer.

The sun climbs quickly, and so does the temperature. Once the beautiful tigress' body temperature reaches the low 30s, she begins to run around. It is late summer, and she is hungry. Any movement catches her attention. With bulging compound eyes, she scans the open ground for prey. Then, at incredible speed, she rushes forward. Whoa! Hard to see when you are running that fast! She slams on the brakes for a fraction of a second, gets her bearings, and redirects the charge. This time, the intended victim turns out to be nothing but a dark plant seed rolling down the slip face of the dune. It grates against her long, toothy jaws. The next target, however, is a wandering cutworm, and her mandibles sink deep.

Methodically, rhythmically, the beetle chews. The tips of her jaws slice through the caterpillar's tissues, while deep within the tigress' mouth, the bases of the jaws pulp and mash. Saliva flows freely into this oral mill, and the caterpillar is partly digested before it is swallowed. Oddly enough, the beetle's actual throat is tiny, and all she can swallow is fluids. Calmly, she drinks the caterpillar slurry.

Then she runs away, and wow, this gal can really run. Oblique tiger beetles have been clocked at 0.42 metres—about 29 body lengths—per second. If grizzly bears could do that, they could charge at 220 kilometres per hour. There are biomechanical reasons why this is impossible, but still, at the scale of a tiger beetle, that is indeed what the charge would look like. (There is, by the way, a flightless Australian tiger beetle that can run 170 body lengths per second. If a grizzly could do that, it could run faster than the speed of sound!)

Fall is on the way, so the beetle needs to fatten herself up. She will spend the day searching for small creatures to kill and eat, and she may also chew on the occasional dead insect or the remains of a small, fallen vertebrate. At around 7 or 8 o'clock in the evening she will dig another overnight burrow. If a storm develops in the afternoon, she will sense the rain coming and dig a burrow for shelter, with ten minutes or so to spare before the first raindrops fall.

Temperature is all-important to tiger beetles. Too cool and they cannot remain active; too hot and they risk death. For most, a body temperature of about 50°C is fatal, and a dry surface on a hot summer afternoon can easily reach that and more. So the beetles regulate their body temperature with their behaviour. In the morning they bask, hugging close to the ground. As the temperature rises, so do the beetles—up onto their tiptoes to keep the body as far off the hot ground as possible. When that no longer does the trick, they duck into the shade for a few moments, and then run back into the sun. It's like you or me running barefoot on a hot beach, from the shade of one palm tree to the next.

Farther south, tiger beetles live in fear of lizards, but not here in Alberta. Our only lizard is the short-horned, and it eats almost nothing but ants. Here, the tiger's enemies are mainly birds and robber flies, and most days the birds are done eating by the time the tigers come out to play. It is the robber flies they would fear, if they had any idea what fear was.

Back to our beetle. She rushes about on the sand and spooks a nearby blowout tiger into flight. As the flying tiger gains altitude, a sleeker, faster insect rockets up from the ground and intercepts it in mid air. Sure enough, it is a robber fly, and its long hairy legs grip the beetle from above while the mouthparts pierce the abdomen. Where the spread wing covers expose thin cuticle, the tiger is briefly vulnerable while in flight. Together, beetle and fly fall to the ground, where the predator sucks the life from its prey.

Sometime in September or early October, our beetle enjoys her last field day of the season. She spend most of it digging an especially long burrow (30 cm—a foot—or more in length), leaving an especially large pile of sand at the entrance. At the end of the day, she retreats to the bottom of the tunnel, and the next morning's breeze covers the entrance with drifting sand. When winter sets in, she is far enough down to escape the frost, but even if the ground around her freezes solid, she will survive. In these environments, snow cover is a great insulator. Some winters are snow free, however, and then the cold penetrates deep.

Winter passes, the days lengthen, and once again the beautiful tigress feels the call of warmth and sun. She emerges in late March, to a dry-looking landscape with none of the still-green growth that graced the dunes in the fall. Still, the sand is hot, and there is much to eat. Within her body, the cooling period has triggered a change in her reproductive organs, and as the days pass, her ovaries swell and fill with developing eggs. For the nearby males, the same sorts of swellings take place in their testes.

Now, foraging takes on another dimension. Male beetles still hunt food, but they will also attempt to mate with almost any other tiger beetle they see. Our beautiful tigress soon learns the rules of this game, as a male rushes up behind her and in a flash has her thorax in his jaws. There is a groove in the

(top) Sticking its head in the sand, a twelve-spot tiger beetle digs an overnight burrow.
(bottom) A big, mean, fast, nasty robber fly, the terror of the tiger-beetle world.

side of her thorax (the COUPLING SULCUS), and it fits his mandibles exactly. If he had tackled a male, or a female of another species, the groove would not have fit, and he would have dismounted. As it is, he attempts to copulate with the tigress, while spreading his forelegs out wide in the air. Some beetle people have suggested that the extra hairs on the front toes of the males are adhesive and are used in mating, but you watch and see: they always hold them out to the sides.

Something doesn't please the tigress. Perhaps the male is too small, or not strong enough, or smells bad, or has a lousy personality (we don't yet understand what makes a tiger beetle a stud or a dud). She kicks at him, rolls on her back, and dislodges him. Moments later, another male tries, but this time she allows him to mate. Antennae twitch, bodies pulse, and it is pretty obvious what is taking place, even across the gap that separates human from insect. After copulation, he rides around on the female's back for some time, just to prevent other males from mating with her as well.

The male's sperm does not go directly to the unfertilized egg. Instead, it remains in a sock-shaped structure within the female until it is needed, just before the egg is laid. Among tiger beetles, the eggs are fertilized according to the "last sperm in, first sperm out" rule, so the male doesn't want competition before she lays her eggs.

With the tip of her abdomen telescoped out to an astonishing length, the female lays one egg at a time, digging a tiny hole for each one, up to a centimetre deep (a little less than half an inch). While she does this, she adopts an odd posture, with her head held high and her body almost vertical. Then she smoothes the surface, hides the hole, and leaves the egg to hatch. To me, the egg is the most amazing survivor in the entire tiger-beetle life-cycle. After all, when the sand gets too hot, the adult can take refuge in the

Sex on the sand: two beautiful tigers of the gibsoni *race, from Saskatchewan.*

(hole): The circular burrow of a tiger-beetle larva, with excavated sand nearby. (head): The head and pronotum of a larva in wait for prey. (larva): The weird-looking larva in its burrow, against the glass wall of a terrarium.

shade, the larva can drop down its burrow, but the egg can only wait and tough it out. A female tiger beetle, by the way, can lay over 100 eggs in her lifetime. But usually she either dies of old age or is eaten by a robber fly or other predatory creature sometime before that.

When the egg hatches, the young larva digs a vertical burrow and begins to wait patiently at the top. It rests with its head flush with the ground surface, jaws ready to strike. When another insect passes by, the lightning-fast

(top) A bee fly—probably a wasp parasite but much like those that attack tiger-beetle larvae.
(bottom) An ant-like Methoca *wasp searching for tiger-beetle larvae to parasitize, on a clump of dried-out mud.*

larva lurches from the burrow and grabs the victim. Then it pulls its meal down into the ground to be consumed. Those insects that resist and try to pull the larva out of the sand are in for a surprise. On the top of the fifth segment on the larva's abdomen are two stout hooks. These dig into the side of the burrow, holding the larva tightly in place. (Flickers, however, have been known to eat tiger-beetle larvae—catching them right at the surface with a quick jab of the woodpecker bill.)

While robber flies stalk the adults, another assortment of insects is searching out the larvae. Predators on the larvae appear to be few, but the omnipresent hister beetles that live in the sandy haunts of the tiger have been known to attack larvae underground.

No, it is parasites that are the larva's biggest worry. *Methoca,* a type of solitary wasp, specializes in parasitizing tiger-beetle larvae, and the wingless females of this insect can often be seen while you hunt for the beetles. The bee fly *Anthrax* is known to flick eggs into larval holes, whereupon the bee-fly maggots enter the larva's body, eventually killing it by devouring its essential organs. Most of the bee flies you see will be parasites of digger bees and wasps, mind you, but for many species we don't yet know what the fly's favourite targets really are. They like to flick eggs into small holes, and a bee-fly specialist, Niel Evenhuis, tells me they will even try to parasitize your shoes by flicking eggs in the lace holes! (Don't worry, the baby bee flies are harmless to people.)

The tiger-beetle larva goes through three INSTARS, between which it moults its entire exoskeletal skin. There are two life-history patterns among the Alberta tiger beetles. First, there is what we call the "spring-fall adult" life-history pattern. A typical example is the bronzed tiger beetle, in which the first instar takes about two weeks and the second takes three. The third instar overwinters. The following summer, it moults to form a pupa (the resting stage, equivalent to a butterfly chrysalis), emerging a few weeks later as an adult beetle. Second, there is the "summer adult" life-history pattern, in which adults emerge in early summer, become reproductively mature without delay, and die off before the fall, leaving only larvae to overwinter.

Although these two groups differ greatly in their winter biology, adults of both groups share almost identical day-to-day lives. Tiger beetles in Alberta have evolved a simple, successful way of doing things, and they all go about their business in more or less the same general way.

As tiger-beetle habitats go, this blowout near Fort McKay is comparatively lush.

2 Why Would They Want to Live *There?*

MOST TIGER BEETLES LIVE IN BARREN PLACES. They do so for one simple, obvious reason: if you find your food by sight, and chase it down at high speed, you shouldn't live in places with a lot of plants. Plants get in the way.

Of course, there is one important consequence of living in these sorts of places, and that is the lack of shade. Things get mighty hot on a sunny day down at ground level, and tiger beetles have to deal with the heat. They also have to deal with their own conspicuousness and the threat of attack by larger predators with equally sharp eyes and speedy reflexes.

Open-ground habitats come in a predictable variety of forms. In sandy country, wind can create either large blowing dunes or small "blowouts," depending on the extent of the disturbance that starts the formation of the habitat. In Alberta, such places are usually the remains of ancient deltas or beaches, dating from just after the last ice age. These habitats are most extensive in the prairie region, but dunes also occur in other parts of the province. The surface of open sand is characteristically hot and unstable, and to an animal the size of a tiger beetle the shifting sands must feel like a wind-blown gravel storm. Compare a freshly emerged tiger beetle, with its luxuriant white pelage, to an old one that has had almost all of the fur knocked off its body, and imagine the forces required to do that. If you and I had hair of equivalent thickness, it would be like coat-hanger wire.

Almost all sandy areas in western Canada are in the process of being overgrown by vegetation. Why, no one seems to know. After much thought, my own opinion is that fire and overgrazing are both relatively unimportant to dune formation, despite often-repeated claims to the contrary. Instead, I believe that drought is the main cause of blowing sand, and I don't think recent "droughts" have been severe enough to keep many of our dunes open and blowing The dust-bowl era of the 1920s and '30s seems to have had a much greater effect on dune formation in western Canada. As well, anyone who spends time in these environments will realize that once the habitat is available to sand-loving animals, the creatures themselves bring literally tons of sand to the surface every year. When you add up the sand from every

The open face of a prairie sand dune is home to only sandy and ghost tiger beetles.

pocket gopher, kangaroo rat, tiger beetle, digger wasp, and ant colony burrow, the combined effects may well be important indeed. More loose sand means more dune movement, which in turn discourages plant growth.

Salt flats are another great habitat for tiger beetles. These areas are formed not by wind erosion, but rather by the actions of water. Salt flats are not as warm as sandy spots, even when they are dry, but they are chemically hostile and thus support little in the way of plant growth. You will hear some people refer to them as "alkali flats" from time to time, but while the surface salt deposits may be alkaline, the underlying soil is often acidic, at least at some depths. "Salt flats" is the term I prefer.

The margins of lakes and streams are also attractive to some sorts of tiger beetles, and they remain free of vegetation because of the actions of water and ice. These habitats include beaches, dry lakebeds, gravel bars and sandbars, mud flats, and erosional cut banks. When they are moist, they are much less hostile than hot open sands. Of course, when they are under water, they are worse than any upland habitat could possibly be. Some tiger-beetle larvae can remain in their burrows for a week or more under water, switching from oxygen-based metabolism to anaerobic body chemistry. Still, many must die when their burrows are washed away by floods. The adults, of course, simply flee the rising water.

(top) Rancher's wasteland, entomologist's wonderland: salt flats near Jenner, home to two rare tiger beetles.
(bottom) Dried salty crust on the pitiful vegetation at the Jenner salt flats.

Sun, lakeside sand, and a few scraggly grasses: a beach tiger beetle's idea of paradise.

Other sorts of upland habitats include clay banks and bare hillsides, usually created by landslides. Sometimes, these hillsides are associated with river channels, although the water that eats away at the bottom of the hill may be a long way down from the high, dry clay banks on which the beetles are found. Temporary stream channels can also create cut banks without forcing the beetles to live in proximity to permanent water.

Finally, there are the semi-open habitats that some prairie species seem to prefer, where plants—mostly short grasses—are indeed growing but with plenty of open space between them. In these places, it is the overall harshness of the dry, sun-baked land rather than geological erosion that makes the habitats usable to beetles and unpleasant to plants.

Habitats that are created by the wear and tear of erosion are often short lived. Beaches come and go, dunes are overgrown by plants, steep hillsides slump downward and become grassy slopes, and sandbars are reshaped each spring by runoff. Tiger beetles have to be able to find these places soon after they have been created, and they have to be able to pull up stakes once the neighbourhood goes to the dogs. Tiger beetles are capable of producing large populations fairly quickly, but their numbers can plummet just as fast.

To illustrate, let me relate the story of Aspen Beach, on Gull Lake, one of my favourite spots for nature study. When my grandmother was young, the lake came right up to the cabins in what is now the summer village. Changes in drainage patterns have affected the lake level since then, and now the shoreline is a few hundred metres away. The wide, sandy beach is only partly natural, since it is kept free of weeds (i.e., unwanted, albeit native, plants) by the village tractor. Every year, the beach and all of the paths leading to the cabins have to be plowed, to keep the plants from overgrowing them.

In the early 1970s, the lake level was dropping especially fast, and the plowed sand on the beach became dry. It then began to drift with the prevailing wind, and a ridge of dunes formed just up from the plowed area. That was about the time I first discovered tiger beetles. At first, I found only the bronzed, twelve-spot, and beach tigers. Then, as the years went on, oblique and sandy tigers became more and more common. Populations of the first three species were huge, and the place was a tiger-beetler's dream.

Then plants, especially sow thistle and willow, rapidly colonized the dunes. Slowly, the habitat became unusable to the beetles, and now only the Twelve-spot is easy to find, and only in small numbers. All of the open sand is too frequently plowed to allow tiger-beetle larvae to survive, and the good old days have vanished. Such is the boom-and-bust world of the tiger beetle, with plenty of frontier communities and plenty of ghost towns. Each individual patch of open ground has its own story, and its own history of fires, droughts, tornadoes, bison stampedes, or what have you.

Of course, not all tiger-beetle habitats come and go quickly, and a few are like grand old cities. My own work on sand dune tiger beetles suggests that the largest dune fields have probably been home to tiger beetles since the end of the last ice age, some 10,000 or so years ago. I believe that the *athabascensis* race of the beach tiger evolved during this period in the Lake Athabasca sand dunes, while the *nympha* race of the sandy tiger evolved in the Great Sand Hills of Saskatchewan. In a very real way, these places are ecological islands in a sea of uninhabitable greenery.

The muddy, messy, but tiger beetle-rich banks of the North Saskatchewan River in Edmonton.

Features used to identify tiger beetles, shown on a blue-morph blowout tiger.

3
The Tiger Beetles of Alberta

THE FOLLOWING ACCOUNTS treat all of the tiger beetles species ever found in Alberta. Each one begins with a descriptive haiku, just for fun. Then, I have tried to simplify identification by presenting field marks in one section and features best seen with a beetle in the hand in another. The labelled photograph opposite shows the various features used to tell different sorts of tiger beetles from one another. "Length" refers to the distance from the front of the eyes to the tips of the wing covers. Males and females generally look alike, although females are a bit larger and have less-hairy front foot toes (TARSOMERES) than the males.

I discuss the classification of each species in three sections: the species' name, its geographic races, and its colour morphs. I present a suggested pronunciation for the scientific name, along with a translation. I also give each species an English name, since most have never had one. Where an existing English name made good sense, I used it—otherwise I made up my own.

Pronunciation of Latinized scientific names adheres to no single standard, so I have based my suggestions here on the most common pronunciations I hear among my friends and colleagues. Admittedly, there are systems for pronouncing Latin, such as the one used by the Catholic Church or the one used to teach Latin to English-speaking school children. None of these has been accepted as a standard by biologists, and since the different systems suggest different pronunciations, I have chosen to avoid the confusion they create. The word *Cicindela*, for example, could "officially" be pronounced "kye-KINN-dell-ah" or "siss-inn-DAY-lah" but since no one I know ever uses either of these options, I will stick to the familiar "siss-inn-DELL-ah." After all, I don't want my readers sounding peculiar in entomological circles.

Geographic races, or simply "races," are equivalent to the biologist's concept of subspecies. I think of a subspecies as a population or populations occurring in a particular geographic area, in which at least 75% of the individuals are distinguishable from those of other subspecies. When they meet, members of different subspecies can and do interbreed, producing intermediate populations, often called "intergrades." As for colour morphs, these are simply variations within the races and are of no more taxonomic importance than people with differently coloured eyes or hair. Still, they make tiger-beetle-watching (like people-watching) more interesting, so I have discussed them in some detail.

I arranged ecological information in three sections. The habitat section gives information on the landscape features that are obvious to people in the places where you find each species. The life history is briefly summarized. Then, the species' geographic range in Alberta is described. Be aware that beetles live only in appropriate sites within the shaded regions shown on the maps. For each species, a more-or-less guaranteed viewing location is mentioned.

The map on the next page shows my own favourite tiger-beetle hot spots (the most accessible locations where good numbers may be encountered) mentioned in the text, as well as the so-called natural regions of Alberta. These are based on features of topography and vegetation, and provide a useful shorthand way to discuss the ranges of living things in Alberta.

To close each treatment, I have added a section called "Cicindelobilia." Here, I have reported personal anecdotes, stories from other researchers, and anything else I thought might increase your appreciation of each and every unique tiger beetle species that lives here in the province of Alberta.

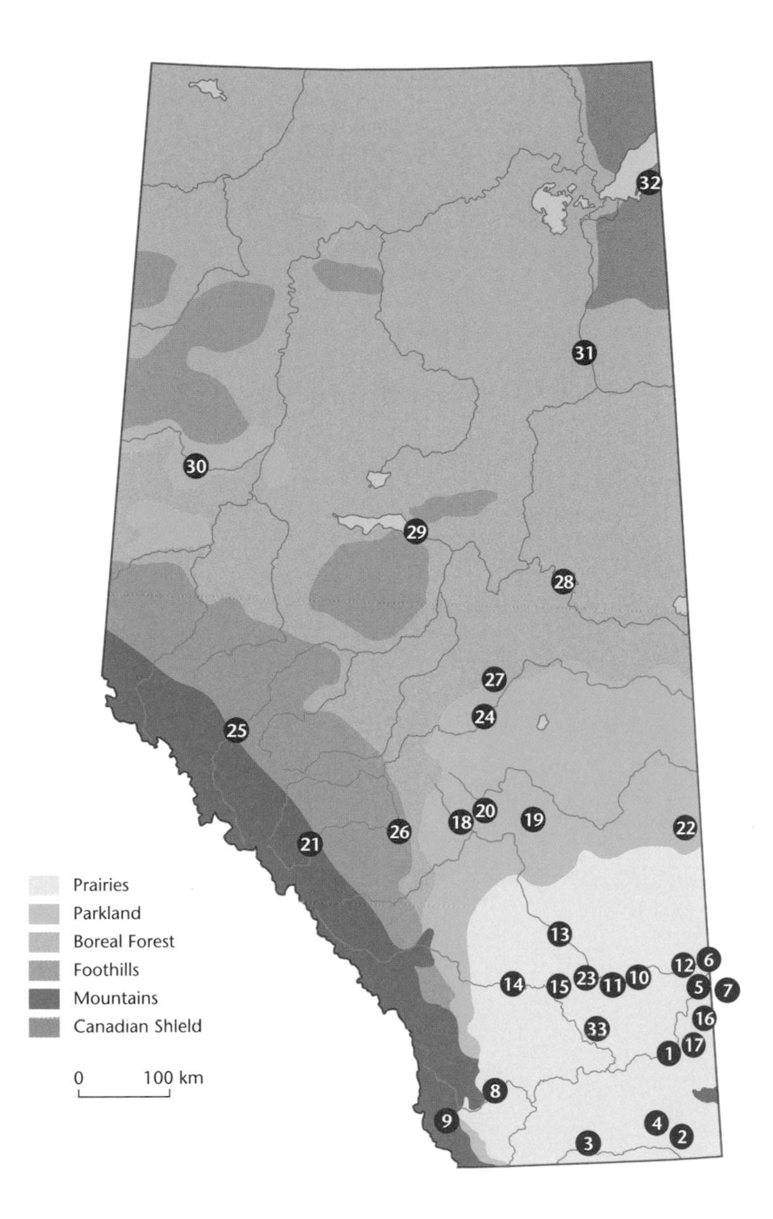

1. Medicine Hat
2. Onefour
3. Milk River
4. Pakowki Lake Sand Dunes
5. Sandy Point Salt Flats
6. Empress
7. Burstall, Saskatchewan
8. Fort Macleod
9. Beaver Mines Lake
10. Jenner Bridge Salt Flats
11. Dinosaur Provincial Park
12. Bindloss Sand Hills
13. Drumheller
14. Calgary
15. Bassano
16. Hilda Sand Hills
17. Chappice Lake
18. Gull Lake
19. Buffalo Lake
20. Wolf Creek Sand Hills
21. Kootenay Plains
22. Provost
23. Gem Sand Hills
24. Edmonton
25. Brule Lake Sand Dunes
26. Rocky Mountain House
27. Opal Sand Hills
28. Lac La Biche
29. Lesser Slave Provincial Park
30. Dunvegan Provincial Park
31. Fort McKay
32. Lake Athabasca Sand Dunes
33. Rolling Hills Sand Blowouts

Cicindela cinctipennis, *our smallest and cutest tiger beetle.*

Tiny Tiger Beetles

ALFRIED VOGLER, at the Natural History Museum in London, England, believes that this is the most primitive group within the genus *Cicindela*. In other words, these beetles are much like the long-extinct common ancestor of *Cicindela* as a whole. In contrast, the traditional entomological arrangement considers them quite highly evolved compared to the ancestral type. In this book, I am placing the tiny tiger beetles first, if only to encourage new ways of looking at tiger-beetle relationships.

The name *Cylindera* means a "cylinder" or "roller," although the relevance of this is even less apparent than "glow worm" for *Cicindela*. I suppose the body of this beetle is somewhat cylindrical, but then so is mine. We have only one Albertan species in this group, *Cicindela* (*Cylindera*) *cinctipennis*. (Note that we write the subgenus name in parenthesis, after the genus name, when writing the name in full.) These "tiny tiger beetles" (my own name for this subgenus) are generally small-bodied and appear as adults in mid summer, wintering as larvae.

An especially brown male grass-runner tiger beetle, in the Tolman Bridge badlands east of Trochu.

grass-runner tiger beetle

CICINDELA CINCTIPENNIS ("Siss-inn-DELL-ah SINK-tih-PENN-iss")

Running all around,
In a garden, of all places!
What does a beetle care?

IDENTIFICATION

In the field: Our smallest tiger beetle; greenish brown with thin, pointy white markings on the wing covers. The shoulder mark and middle band are separated from the wing-cover edge by a dark margin, which is narrow in the *cinctipennis* race and broad in the *imperfecta* race.

In the hand: Forehead not hairy, underside of body iridescent green with white hairs lying flat against the surface.

Length: 8–10 mm.

Similar species: Size alone makes these beetles tough to mistake for any other species, and their markings are also unique.

THE NAME

Since it lives in grassy places and prefers to run rather than fly, I have called this beetle the "grass-runner." The word *cinctipennis* means "surrounding feathers." Believe it or not, the species name refers to the light markings around the outside edges of the wing covers, since insect wings are often referred to as "pennis" in scientific names. The name *imperfecta* refers to an incompleteness of some sort, possibly the separation of the wing-cover markings from the margin. *Cicindela pusilla* and *Cicindela terricola* are names you may see used elsewhere for this species. The former is now considered invalid, and the latter most likely refers to a separate, eastern species, although opinion is still divided on this issue. *If C. cinctipennis* and *C. terricola* are eventually determined to be two names for the same species, the name *C. terricola* will become the correct one for our Alberta beetles.

CLASSIFICATION

Geographic races: The *cinctipennis* race is widespread in the prairies east of the Rocky Mountains. It is a typical-looking tiger beetle, but in miniature. West of the Rockies, and historically in Alberta's Kootenay Plains, one finds the *imperfecta* race. It is a wee bit bigger and more brilliantly green on the underside, and its wing-cover markings are separated from the edge by a broad dark border. This gives it a pattern that is quite unusual for tiger beetles in this area, but shared by other tiny tiger beetles farther south. The southwestern tiger beetle *C. lemniscata* is even tinier and has been interpreted as an ant-mimic. Supposedly, its white markings look like the outline of a copper-coloured ant. Whether this can explain the pattern on our *imperfecta* is unknown, and to me it seems unlikely.

Colour morphs: As far as I have seen, this species does not exhibit colour morphs in Alberta, although there is some variation from greenish brown to brownish green ground colour.

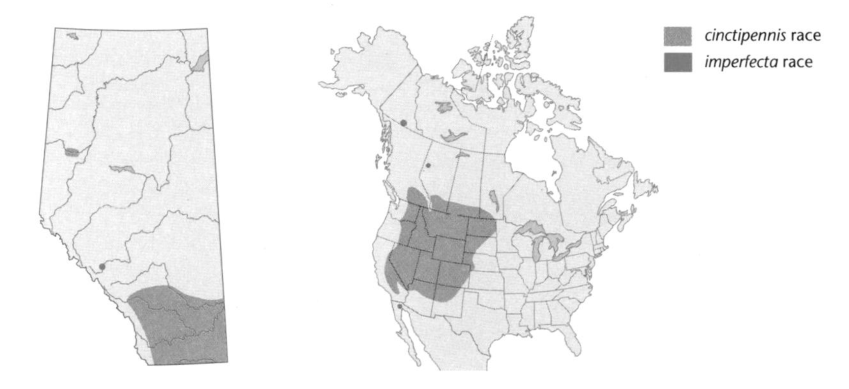

ECOLOGY

Habitat: This is a species of sparsely vegetated areas in the prairies. It seems to be most abundant in saline places and on the flats at the bases of badland hills, but I have also seen it in cultivated gardens in Medicine Hat.

Life history: This is a summer species, meaning that it emerges from the pupa sometime in July, mates, and dies sometime in August. The larvae overwinter, but the length of the larval stage in Alberta is unknown.

Range: Grass-runners of the *cinctipennis* race are widespread in the prairies. There is also an isolated population in the Peace River area. The *imperfecta* race is known only from the Kootenay Plains, southwest of Nordegg. Rick

Head-on view of a grass-runner tiger beetle, looking every bit as fierce as its larger relatives.

Freitag discovered a population there in the mid 1960s, in a spot that is now under water, where the Bighorn Dam formed Abraham Lake. We don't know whether they still exist in the province.

Where to find them: The easiest places to see these beetles are in the badlands around Drumheller or in Dinosaur Provincial Park. If you rediscover the *imperfecta* race, let me know so I can see it too!

CICINDELOBILIA

I apologize for the lack of a photograph to illustrate the imperfecta *race. I couldn't find one of these beetles in Alberta, so I called my friend Dick Cannings in Naramata, BC. He was only too happy to look for them and went out the first chance he got, to a spot near the Yacht Club in Penticton. As he entered the appropriate clearing, net in hand, a man happened along the trail and said casually, "so, you're looking for tiger beetles, I assume." Amazed, Dick confirmed that yes, in fact he was, at which point the man went on to say that he had heard all about tiger beetles on television, the night before, on one of my* Acorn, The Nature Nut *shows. When Dick said he was doing a favour for me, this too seemed obvious to the guy. Who else would want tiger beetles? Then they discovered they had another mutual friend in common, who had just called Dick a few minutes before he went out beetling, after many months of silence. But again, the mystery man was unimpressed by the coincidence. Mark my words: these are the sorts of weird things that happen once you become a cicindelophile. And no, Dick did not find the beetles. Another interesting story about this species dates back to the early 1970s, when my friend Gerry Hilchie was a teenager, working as a lifeguard at the Bow View Swimming Pool in Calgary. From time to time, his attention was diverted from the safety of his fellow human beings when he would see a grass-runner on the pool deck, about half a kilometre from any obvious suitable habitat. I suppose a good lifeguard should notice everything.*

Good habitats for the grass-runner, badlands, and cowpath tiger beetles can be found in the river valley at Drumheller.

The ghost tiger beetle, our rarest and most exotic species.

Curlicue Tiger Beetles

THE CURLICUE TIGER BEETLES belong to a single large branch of the family tree of tiger beetles, and are not very closely related to the other species in Alberta. These are summer-adult beetles with delicate, medium-sized bodies, hairs that lie flat against the surface, and very curlicue-ish, whorled markings on the wing covers, sometimes looking like the work of a skilled scroller. They typically live in salty places (the ghost tiger is an exception), and most are not terribly colourful. The name *Ellipsoptera*, by the way, means "defective wing" and probably refers to the slight indentation in the hind edge of the wing covers (which are modified front wings) of some species.

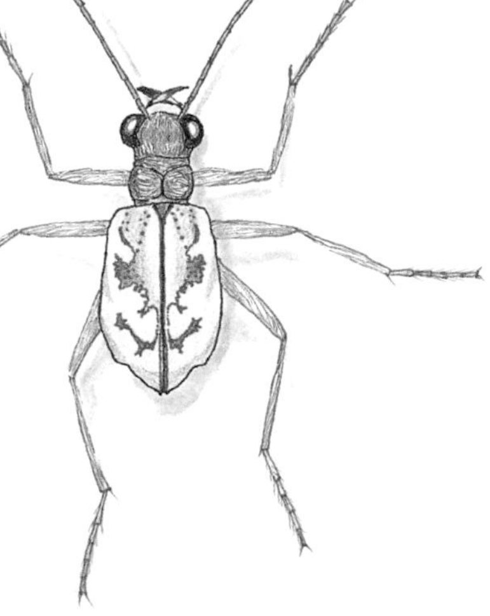

Female salt creek tiger beetle on a moist salt flat near the Jenner Bridge.

salt creek tiger beetle

CICINDELA NEVADICA ("Siss-inn-DELL-ah neh-VADD-ick-ah")

A pinned specimen,
See the look on its dried-out face,
As if it was alive.

IDENTIFICATION

In the field: A medium-sized, relatively dull-coloured tiger beetle with extensive, extra-wiggly pale markings on the wing covers, including a coathook-shaped shoulder mark.

In the hand: Forehead hairy, body hairs lying flat rather than standing up, underside of body only moderately iridescent, blue-green and coppery.

Length: 11–12.5 mm.

Similar species: The beach tiger beetle (*C. hirticollis*) has a similar wing-cover pattern, but is bigger, does not live in salty places, and has upright hairs rather than flattened ones.

THE NAME

This species name is an easy one, referring to the state of Nevada. The English name is a good one too, and it originated in Nebraska, where the *lincolniana* race (named for the city of Lincoln) is threatened with extinction.

CLASSIFICATION

Geographic races: All of our Alberta populations are members of the *knausii* race, named by Charles Leng for W. Knaus, another tiger-beetle specialist.

Colour morphs: There is only one colour morph of this species in Alberta, and all of our salt creek tigers look much like the one in the photograph opposite.

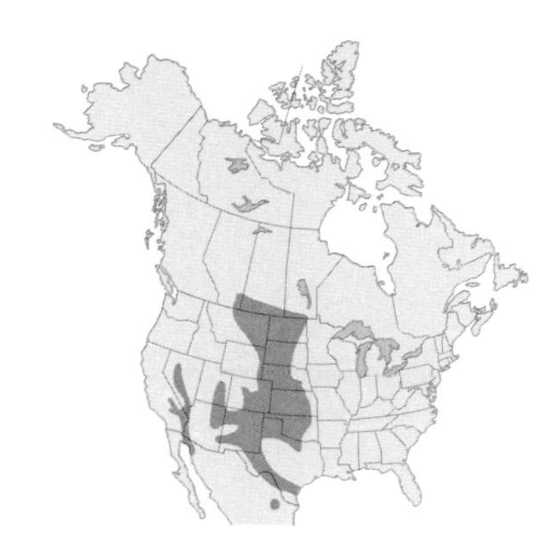

ECOLOGY

Habitat: This is a salt-flat species, but it seems to need wet, muddy spots, rather than dry, crusty ones. I have seen it only in salt flats in the valleys of prairie rivers, moistened by spring-fed seeps.

Life history: Another summer-adult species that winters as a larva. The length of the larval period in Alberta is unknown.

Range: In Alberta, found only in the prairie region.

Where to find them: The best spot I know of in Alberta is just east of the Jenner Bridge. A road on the south side of the Red Deer River leads to a campground, and along the way there are white, salty flats. Look in the wetter areas for this species.

CICINDELOBILIA

Since the publication of J.B. Wallis' book, two tiger beetles have been added to the Alberta list. Both are summer-adult species, easy to miss because they are present for only a short season. The first was the ghost tiger beetle, discovered on the Empress dunes by David Larson, and the second was the salt creek tiger beetle. On the sunny afternoon of July 6, 1974, Gerry Hilchie, Bob Davidson, and I were looking for shiny tiger beetles at the Jenner site. We found them and rejoiced, and then we all caught a few of another species, unknown to us. It turned out to be the salt creek tiger, and we were all happy to be the ones to discover a new species for the province. To get the photograph opposite, Christina Hall and I drove to the very same spot, 24 years later almost to the day. Not only were the beetles still present, but we found them only on the very same wet area, no bigger than an average bedroom.

Up close, the ghost tiger beetle is easy to see, but at a distance it vanishes against the sand.

ghost tiger beetle

CICINDELA LEPIDA ("Siss-inn-DELL-ah LEPP-pih-dah")

Warm prairie night,
Kangaroo rats in the dark,
And there, lepida!

IDENTIFICATION

In the field: A small, light-coloured tiger beetle with faint dark marking on the wing covers, and pale legs and antennae.

In the hand: The white body hairs lie flat against the surface, especially on the forehead, and partially hide the brown and iridescent green underside.

Length: 9–11 mm.

Similar species: Easy to confuse at a distance with the sandy tiger beetle. The vaguely defined dark markings and light legs are good field marks.

THE NAME

Lepida means pretty, neat or graceful. I have called it the ghost tiger beetle since it is often said that the shadow of this beetle is more visible than the beetle itself. In his treatment of the New Jersey tiger beetles, Howard Boyd called this species "the little grey ghost of the dunes," a lovely name that I considered briefly but rejected because of its length.

CLASSIFICATION

Geographic races: Despite being restricted to isolated dune fields, this species is uniform in appearance throughout its extensive range.

Colour morphs: Some are greenish rather than bronzed on the dark parts of the body, but these are downright rare in Alberta. These were once considered a race called *"insomnis"* (meaning sleepless), but now we just call them "greenish."

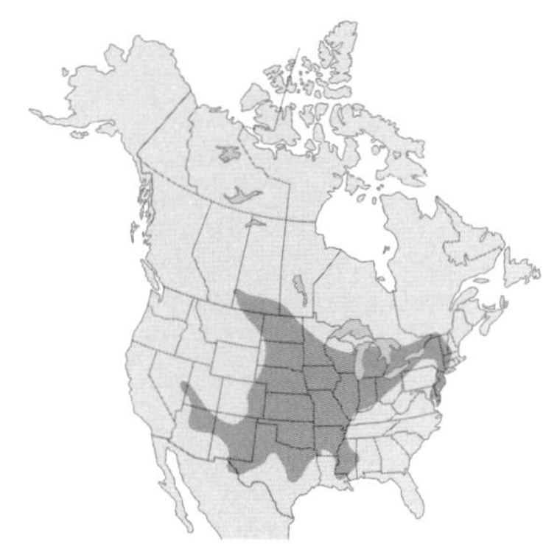

ECOLOGY

Habitat: Open blowing sand dunes, especially near the middle of the dune, and among sparse clusters of scurf pea and other sand-stabilizing plants.

Life history: Ghost tigers appear as adults in the summer, from about mid June to mid August. In Manitoba, Norman Criddle noticed that older adults seem less likely to fly. The larvae take two years to complete development in Manitoba, and probably here as well.

Range: In Alberta, known only from two dune fields. One is approximately eleven kilometres south of Empress and contains only two active dunes. The other is north and east of Bindloss, in a hairpin bend of the Red Deer River valley called "Dune Point," and it also contains two active dunes. The dunes in both locations are gradually being overgrown, and may not persist beyond another decade or two unless something happens to get them moving again.

Where to find them: In Alberta, these beetles live only on private ranch land, in remote localities requiring a four-wheel-drive vehicle for access. The easiest way for an Albertan to see ghost tiger beetles is to drive east into the Great Sand Hills of Saskatchewan. The easily accessible dunes straight south of the town of Sceptre (the same road also runs straight east of Liebenthal) should be swarming with them at the right time of year.

CICINDELOBILIA

At first sight, this beetle seems out of place among the Alberta tiger beetles with its odd wing-cover pattern and pale colours. On closer examination, however, one can see the outline of all the familiar markings, including the shoulder mark, middle band, and tail mark. The pale appendages and dense, white hairs most likely serve to camouflage this beetle and prevent it from overheating in the sun. Unlike most other Alberta tiger beetles, the ghost can remain active after sundown and is often found running on the dunes on warm evenings. As a rule, the tiny, curlicue, and American tiger beetles are the night owls among our species, while the temperate tiger beetles put more emphasis on the value of a good night's sleep. On the Empress dunes, I found that the beautiful tiger beetle can at times be an important predator of ghost tiger beetles. See the sandy tiger beetle account *(pp. 52–54)* *for more details.*

Male backroad tiger beetle, showing off the double row of lovely green pits on his wing covers.

American Tiger Beetles

CICINDELIDIA MEANS "little *Cicindela,*" and that is a pretty good way to think of the members of this group. They are generally smaller than the temperate tiger beetles and not terribly colourful. In the views of Alfried Vogler, at the Natural History Museum in London, they are also the closest living relatives to the temperate tiger beetles. They winter as larvae and are summer-adult species. We have only one representative in Alberta, and it is one of the most abundant and widespread species of tiger beetles in North America. The group lives only in the Americas, hence my choice of English name.

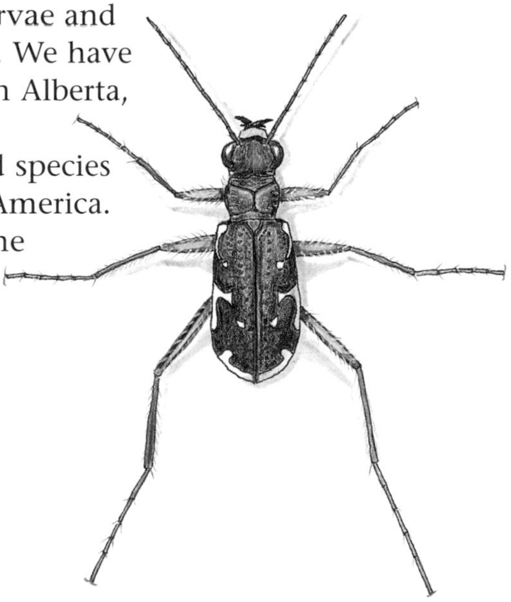

A backroad tiger beetle: slim, athletic, and alert!

backroad tiger beetle

CICINDELA PUNCTULATA ("Siss-inn-DELL-ah PUNK-tue-LAT-ah")

Hot day, dry wind, sand,
Hell of a spot to live in,
But they live with vim!

IDENTIFICATION

In the field: A slender, black tiger beetle with very thin white markings on the wing covers.

In the hand: Blue-green underside on abdomen with purple-coppery patches on sides of the thorax, forehead not hairy, and a row of tiny blue-green pits running the length of each wing cover. For what it's worth, many people have noticed that the defense secretions of this species smell like apples, although this is probably true of most other tiger beetles as well.

Length: 11–13 mm.

Similar species: The black-bellied and cowpath tiger beetles are both sturdier and heavier than the backroad tiger, with very different wing-cover markings, if any.

THE NAME

Backroad tiger beetles often live in exactly that sort of place. The word *punctulata* means "with little pits," and I'll bet this refers to the row of tiny green pits on each wing cover.

CLASSIFICATION

Geographic races: There are no distinctive geographic races of this species in Alberta, and all of our populations belong to the *punctulata* race.

Colour morphs: Although the wing-cover markings vary from non-existent to thin but distinct, there are no obvious variations in ground colour in our Alberta populations. Farther south, a green morph appears here and there (the "Chihuahua" morph), but so far it has not turned up in Alberta.

ECOLOGY

Habitat: This is another species of the sparsely vegetated prairie, although it seems to prefer somewhat sandy soil, and not salty spots. The edges of dirt trails through sandy prairie are perfect places to find them.

Life history: A summer-adult species that winters as a larva. In the United States, the entire life cycle takes only one year, but whether this is possible in southern Alberta, no one yet knows.

Range: A prairie species in Alberta, apparently occurring no farther north than the Red Deer River drainage basin.

Where to find them: The best sites I know of in Alberta are along the sandy road between the town of Empress and the Empress cemetery. The last kilometre or so of the road to the cemetery is especially good.

CICINDELOBILIA

Despite their preference for somewhat sandy soils, I did not find this species associated with sand-dune habitats when I studied the tiger beetles of the Empress area. Instead, they lived on stabilized sand hills that were mostly covered with vegetation. This is an active, wary sort of tiger beetle, and you can tell at a glance that it is slimmer and more of a runner than its burly relatives in the temperate tiger beetle group. I always enjoy seeing backroad tigers, but I rarely see them in great numbers in any one spot.

A shiny tiger beetle enjoying a meal of the universal food for tiger and predatory ground beetles: Purina Dog Chow.

Temperate Tiger Beetles

MOST OF THE TIGER BEETLES in Alberta belong not only to the genus *Cicindela*, but to the subgenus *Cicindela* as well. These are the typical tiger beetles of north temperate regions, and almost all of them overwinter as adults and as half-grown larvae. Until recently, this group was considered quite primitive within the genus (not much modified from the common ancestor). However, molecular studies by Alfried Vogler and his colleagues have suggested that this group may be quite highly evolved, having developed modifications in life history and a large body size as adaptations to cooler climes.

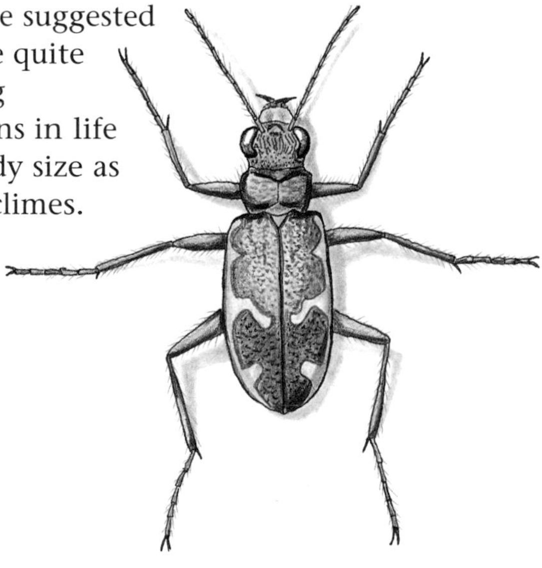

The Sylvan Group

THE TWO SPECIES on pages 37 to 41 are part of the *Cicindela sylvatica* species group, a well-defined subdivision of the temperate tiger beetle group. They take their name from a big, burly European species, and the words *sylvatica* and sylvan both refer to forests, which makes some sense for one of our species, but none for the other. I have placed them first among the following treatments only because they share a bald forehead with their more primitive relatives—a feature that may or may not have much meaning.

A male long-lipped tiger beetle on needle-strewn sand in a jack pine forest.

Female black-bellied tiger beetle from Steveville, photographed in a terrarium.

black-bellied tiger beetle

CICINDELA NEBRASKANA ("Siss-inn-DELL-ah NEH-brass-KAN-nah")

A room in Provost,
Asleep in Don Shaw's closet,
montana *waits outside.*

IDENTIFICATION

In the field: A medium-sized, sparsely marked black tiger beetle (never olive or green), usually with a tan or black (sometimes white) elongate upper lip.

In the hand: Underside of body usually black, rarely iridescent, surface of wing covers smoother than in the long-lipped tiger beetle, especially under high magnification. Also note the non-hairy forehead.

Length: 12–14 mm.

Similar species: Long-lipped tiger beetles are usually iridescent blue, green, or bronzed on the underside, and have larger white markings on the wing covers as well as a white upper lip. Cowpath tiger beetles have hairy foreheads and always have white-tipped wing covers, not black.

THE NAME

This species has been named and renamed many times, but always after places in the west. We called it *Cicindela montana* until Tim Spanton realized that the proper name was *C. nebraskana.* In the early 1900s Colonel Thomas Lincoln Casey confused the issue with the names *Cicindela canadensis* and *C. calgaryana* (the latter, oddly enough, from Lethbridge). Although he contributed much, Casey was notorious for naming fictitious species. A friend of mine at the Smithsonian Institution has lamented this in a song called the "Thomas Casey Blues." I chose the English name "black-bellied tiger beetle" to refer to the best means of distinguishing these beetles from long-lipped tigers.

CLASSIFICATION

Geographic races: Thankfully, this species is not terribly variable, and most of the ones you find will look much like the photograph above. No geographic races are recognized.

Colour morphs: All of the Alberta black-bellied tigers are black, and no other colour morphs are known.

ECOLOGY

Habitat: This is a prairie species that lives in grassy areas. Watch for them in sparsely vegetated places, and even on dried cow pies, but do not expect them on sand or salt flats.

Life history: This is a spring-fall species, with adults emerging in late summer and overwintering. The larvae probably take two years to mature.

Range: In Alberta, this is a prairie species, although some individuals have turned up in the southern foothills and southern parklands as well.

Where to find them: This is not a species one finds in great numbers, but I have seen it often in quite ordinary-looking patches of sparsely vegetated prairie. Thus, I cannot send you to one particular hot spot for these beetles. With a little persistence, however, you will soon run into them in the right habitats.

CICINDELOBILIA

Distinguishing this species from the long-lipped tiger was difficult until my friend Tim Spanton published his Master's thesis, which he wrote at Lakehead University under the supervision of Rick Freitag. It was a very sad day for me when I heard that Tim had passed away in 1993, while only in his 30s. He and I were students together, and I will always be thankful for the time we spent together talking about beetles, birds, science, and the vagaries of life. I did, by the way, see my first black-bellied tiger near Provost, where I shared a motel room with Don Shaw, with my sleeping bag on the floor and my head in the closet.

Long-lipped tiger beetles don't seem to mind foraging in plant debris.

long-lipped tiger beetle

CICINDELA LONGILABRIS ("Siss-inn-DELL-ah LONN-jih-LAY-briss")

What is it doing?
Running on reindeer lichen,
It should know better.

IDENTIFICATION

In the field: A medium-large, black, olive or greenish tiger beetle, usually with sparse white markings on the wing covers and a long, white upper lip.

In the hand: Forehead not hairy, usually with an iridescent underside to the body, and a rougher texture to the surface of the wing covers, compared to the black-bellied tiger, especially under high magnification.

Length: 13–15 mm.

Similar Species: Most black-bellied tiger beetles are black on the underside, have smaller wing-cover markings, and have a dark or tan upper lip. Cowpath tiger beetles have hairy foreheads and always have white-tipped wing covers.

THE NAME

Longilabris means "long-lipped" and refers to the elongate upper lip that this species shares with the black-bellied tiger beetle.

CLASSIFICATION

Geographic races: Over most of the province, long-lipped tiger beetles are black and have small, thin markings. Tim Spanton concluded that these beetles belonged to the widespread *longilabris* race. In the southwestern corner of Alberta, however, this species is extremely variable. Some have no markings, some have heavy markings, some are black, others are olive coloured, and some are bright green. These beetles are apparently the result of interbreeding among three races that were at one point separated. Thus, Tim gave them the ultra-cumbersome scientific name *Cicindela longilabris longilabris* X *Cicindela longilabris perviridis* X *Cicindela longilabris laurentii* (where the Xs mean "crossed with"). If I were you, I would just call them "intergrades."

Colour morphs: Among the *longilabris* race, most beetles are black, but among the intergrades you will find black, olive-green, and bright-green individuals.

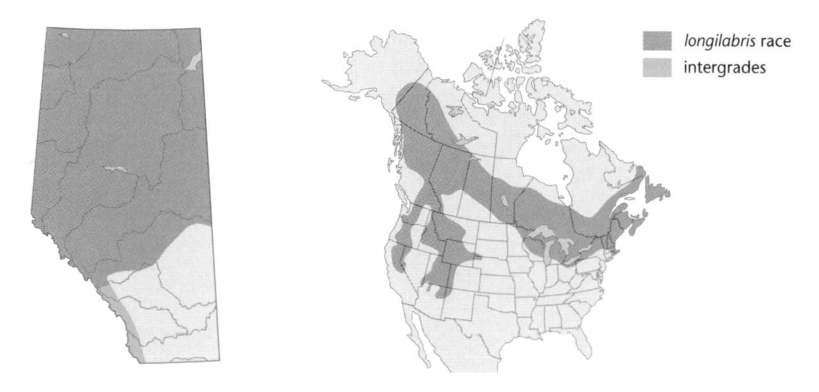

ECOLOGY

Habitat: These beetles can be very common on open or partly vegetated sand, and in forested regions, this usually means areas with pine trees. They can also be found on semi-open ground in mixed spruce and poplar forests.

Life history: Tim Spanton studied the life history of this species in Ontario. There, it lives two years as a larva, emerges as an adult in late summer, overwinters, and breeds the following spring. In the absence of evidence to the contrary, we can assume the same is true here.

Range: This species is found throughout Alberta except on the prairies. Thus, it overlaps with the black-bellied tiger only where the grasslands meet the parklands to the north, and in the montane forests of the southern foothills.

Where to find them: A good spot for the widespread race of this species is the Opal Natural Area. This is an unmarked parcel of crown land that is easily accessible by taking the first left turn north of the town of Opal and continuing until the road forks. Take the left fork and park by the open sand. For those fascinating intergrades, try looking about 17 km south of Beaver Mines Lake.

Bits of beetle-sized pine charcoal mixed in with sand at the Opal Natural Area.

CICINDELOBILIA

In some places, this is a very common tiger beetle, while in others, it is sporadic and rare. At a spot near the zoo in Edmonton, I used to find the long-lipped tiger beetle each year, though only one at a time. It is now a paved bicycle path. In contrast, it is easy to find a dozen in one outing at Opal. These beetles are clearly at home on needle-strewn sand in jack and lodgepole pine forests. In fact, their colour may have something to do with this habitat. Pine trees require fire to open their cones, and as a consequence a healthy pine forest usually has some chunks of old charcoal mixed in with the sandy soil. It is possible, although tricky to prove, that a black beetle is easy to mistake for a fragment of burnt pine bark in these areas. American tiger-beetle specialist Tom Schultz has suggested that resemblance to rocks and other such things is the reason for many instances of puzzling tiger-beetle colouration, and refers to it as "special resemblance." Of course, open sands in pine forests are not as warm as those on the prairie. In such places a dark beetle may have an easier time staying warm when the sun is shining, and this may be the real reason for black colouration. The warmest sand temperature I have measured at Opal was 37°C, a far cry from the 57°C I once encountered on the prairies.

The Maritime Group

IN ALBERTA, five species belong to what tiger-beetle specialists call the *Cicindela maritima* species group, or the "maritime group" in English—a subgroup of the temperate tiger beetle group. They are more closely related to each other than to any other tiger beetles and take their name from a European relative. As you will see from the accounts, they are also similar in their life histories and, to an extent, their habitats as well.

The bronzed tiger beetle, a short-lipped, hairy-headed, very familiar species.

A male bronzed tiger on an especially clean and quartz-rich bit of beach sand at Buffalo Lake.

bronzed tiger beetle

CICINDELA REPANDA ("Siss-inn-DELL-ah ree-PAND-ah")

In the homemade net,
I turn my first one over,
And it bites me hard!

IDENTIFICATION

In the field: A medium-sized, bronze-coloured tiger beetle with a short shoulder mark that connects, or nearly connects, to the other markings along the edge of the wing cover.

In the hand: Hairy forehead, relatively narrow prothorax compared to that of the twelve-spot tiger, underside of abdomen blue-green, with coppery sides to the thorax.

Length: 11–13 mm.

Similar species: On twelve-spot tigers, the shoulder mark is clearly separate from the other light markings, the prothorax is wider, and the beetle's overall colour is darker brown. Beach tigers are larger, have a different body shape (females are broader, males are more elongate), and have coathook-shaped shoulder marks.

THE NAME

Repanda means bent backwards or turned up, presumably referring to the shape of the shoulder mark, or one of the other markings. It could also refer to the French word *répandu*, meaning "widespread." This seems likely, since P.F.M.A. Dejean, a French entomologist, coined the name. I have called it the bronzed tiger beetle to draw attention to its colour, which is the easiest way to separate it in the field from the similar twelve-spot.

CLASSIFICATION

Geographic races: This is not a variable species in Alberta, and most bronzed tiger beetles look much like the one shown opposite. The *repanda* race is the one we find in Alberta.

Colour morphs: All of the bronzed tigers I have seen in Alberta were the usual bronze-brown.

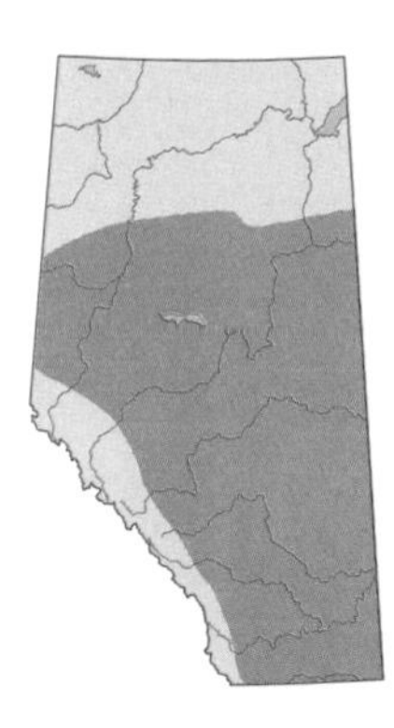

ECOLOGY

Habitat: Found on open mud and wet sand near the edges of lakes, streams, and rivers. These beetles seem to like it right down along the water's edge.

Life history: Adults appear in the late summer or fall, overwinter, and breed the following spring.

Range: The bronzed tiger is widespread in appropriate habitats, but not found commonly in the mountains, and not at all in the far north. It seems to be equally at home near lakes, rivers, and streams, so long as there is appropriate open ground to hunt on and for the larvae to dig their burrows.

Where to find them: On a sunny day, almost any bit of sand or mud along a river or lake should produce at least a few bronzed tigers. If it doesn't, you are probably there at the wrong time of year, as they do disappear from some sites in midsummer.

CICINDELOBILIA

This species achieved some notoriety in biological circles when David Sloan Wilson published a study in which he suggested that adult bronzed tigers were leaving perfectly edible bugs alone so that their nearby larvae would have more to eat. To my knowledge, no one has followed up on this study, but at a gut level I find it pretty hard to believe. If there is such a thing as a selfless tiger beetle, their hard-hearted image will suffer no end. *One of my most interesting moments with this species occurred at Gull Lake, where I saw a bronzed tiger jumping about 20 cm into the air and then running in tight circles. Above it, a small cloud of midges (flies in the family Chironomidae) were engaged in a mating swarm. The beetle was trying to catch them, but as far as I could tell it was completely unable to do so.*

With its head held high, a male twelve-spot tiger beetle shows off its lovely brown wing covers.

twelve-spot tiger beetle

CICINDELA DUODECIMGUTTATA ("Siss-inn-DELL-ah DOO-oh-DESS-im-gut-TAY-tah")

All along the stream,
Mud-coloured beetles with spots,
Like they have no fear.

IDENTIFICATION

In the field: A medium-sized, dark-brown tiger beetle with a short shoulder mark that is clearly separated from the other light markings.

In the hand: Hairy forehead, relatively wide prothorax, underside of body mostly iridescent blue-green, sides of the thorax coppery.

Length: 11-13 mm.

Similar species: Bronzed tigers are lighter in colour, have a narrower prothorax, and bear shoulder marks that connect or nearly connect to the other markings. Pacific tigers have only a few hairs on the forehead, and the shoulder mark usually consists of only two dots. Hybrids between Pacific and twelve-spot tigers are intermediate in their features.

THE NAME

Duodecimguttata means twelve-spotted. Thus, the common name is a direct and appropriate translation, although tiger-beetle experts rarely count spots the way ladybug people do, and prefer instead to pay close attention to the shapes of the spots.

CLASSIFICATION

Geographic races: Some of these beetles have relatively well-marked wing covers, while others, even outside the zone of interbreeding, possess thin, scant markings. Still, there are no discrete geographic races here, or elsewhere for that matter.

Colour morphs: Near the east coast, you can find greenish and even blue colour morphs of this beetle, but here in Alberta they are almost always muddy brown.

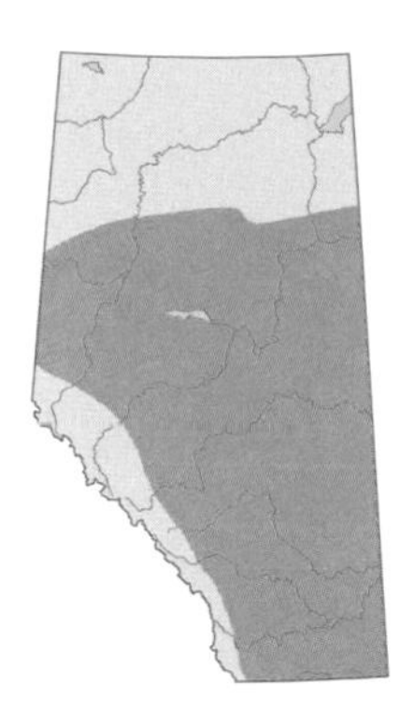

ECOLOGY

Habitat: Found near water, on wet sand, clay, mud, or any mixture of the three, but generally farther from the water than the average bronzed tiger. From time to time, twelve-spots also inhabit dry sand dunes, clay banks, and saline flats. Most of these sightings happen in late summer and may represent beetles that have flown away from their breeding habitats, in search of muddier horizons.

Life history: Adults appear in late summer or fall, overwinter, and breed the following spring.

Range: The twelve-spot is widespread outside the mountains. Rick Freitag showed that there is a zone in the foothills, about 70 km wide (where he measured it west of Rocky Mountain House), in which this species and the Pacific tiger beetle interbreed. Rick is a retired professor at Lakehead University. His work on the maritime group, published in 1965, began his long and fruitful career as a tiger-beetle specialist, which got off to a roaring start at the University of Alberta under the supervision of George Ball.

Where to find them: This species, in my experience, is even easier to find than the bronzed tiger. Any stream, river, lake, or pond with an open shoreline should support at least a few.

CICINDELOBILIA

Admittedly, this tiger beetle does little to draw attention to itself. You will find, however, that you get a sense of accomplishment from learning to distinguish it from the bronzed tiger beetle. Finding a similar but different tiger right alongside the bronzed can be magically appealing to those of us who truly enjoy biodiversity.

With the sun setting slowly near Brule Lake, a male Pacific tiger beetle hugs the ground for a last bit of warmth.

Pacific tiger beetle

CICINDELA OREGONA ("Siss-inn-DELL-ah ORE-egg-OWN-ah")

Look—oregona!
Hmm, I didn't expect it,
Trout aren't biting now.

IDENTIFICATION

In the field: A medium-sized, dark-brown tiger beetle with meagre light markings on the wing covers, usually reduced to dots rather than forming crescents or lines.

In the hand: Only a few hairs above the eyes on the forehead, relatively broad prothorax, underside all iridescent bluish or with coppery sides to the prothorax.

Length: 11–13 mm.

Similar species: This species is most similar to the twelve-spot tiger, and is best distinguished by its balding, but not quite hairless, forehead. The wing-cover markings are so small that there is little chance you will confuse it with the bronzed tiger, and it differs from that species in all the same ways as the twelve-spot. As you might expect, hybrids between this species and the twelve-spot are intermediate in their appearance.

THE NAME

Oregona refers to the state of Oregon, but these beetles are no more typical of Oregon than anywhere else they occur. I have called the species the Pacific tiger beetle, a name that reminds us of its western roots.

CLASSIFICATION

Geographic races: On the underside of the thorax, most of our Pacific tigers are coppery like the *guttifera* race, which is common along the east slopes farther south. However, in our area they interbreed with the race *oregona* (with blue on the underside). Therefore, most of our beetles should be considered intergrades. These beetles also interbreed with the twelve-spot tiger along the foothills.

Colour morphs: Elsewhere, these beetles can be blue or green overall, but here in Alberta they stick to a nice, conservative brown.

ECOLOGY

Habitat: Since they live alongside water in the mountains, these beetles often find themselves on gravel and coarse beaches. After all, mountain streams have gravelly bottoms while slower prairie streams, with less ability to carry sediments, are muddier. However, I have found Pacific tigers on sand at the Brule Lake sand dunes just east of Jasper National Park.

Life history: Like other temperate tiger beetles, this species emerges from the pupa in late summer, overwinters as an adult, and breeds the following spring.

Range: In Alberta, restricted to the mountains and foothills.

Where to find them: I started seeing this species on a regular basis only when I became a fly fisherman. Moving slowly along streams in the mountains and foothills, and using gravel bars as kneeling spots while casting to rising trout, I often spooked a Pacific tiger or two.

CICINDELOBILIA

Once he realized they were interbreeding, why did Rick Freitag continue to recognize both this species and the twelve-spot, rather than lump them into a single species? The situation is almost exactly like that with orioles. Do you remember when we had a bird here called the Baltimore oriole, distinct from the Bullock's oriole to the south? Then ornithologists discovered that they were interbreeding and decided to lump them both into the northern oriole. Bird watchers and sports fans were downcast. Now, however, the two original names are back, since recent studies concluded that even though the two birds could interbreed, they were not in the process of blending together. The hybrid zone is stable, and outside of it the two species are doing their own thing. And that is exactly what Rick Freitag thought when he studied these two tiger beetles.

The clean sand at Buffalo Lake is a perfect substrate for this big, bronzy female beach tiger beetle.

beach tiger beetle

CICINDELA HIRTICOLLIS ("Siss-inn-DELL-ah HURT-ih-KOLE-iss")

Twelve miles of sand dune,
And here, where no birds fly,
Old hirticollis!

IDENTIFICATION

In the field: In the southern half of Alberta, these beetles look a lot like the bronzed tiger, but with a coathook-shaped shoulder mark and a slightly bigger, hairier body. Females are noticeably broader in the beam, while males are longer and more straight-sided than bronzed tigers. The northern *athabascensis* race is distinctive, with a reduced shoulder mark, and about one-half of the population bright green or blue.

In the hand: Hairy forehead, blue-green underside of abdomen, sides of thorax coppery.

Length: 11–15 mm.

Similar species: The bronzed tiger is smaller and has a different shape to the shoulder mark. The salt creek tiger has similar markings but has hairs that lie flat rather than standing on end, as well as being smaller and living in a different habitat.

THE NAME

The word *hirticollis* means hairy-necked and is appropriate for this species, although many other tiger beetles have hairy necks. I chose "beach tiger beetle" to refer to the usual habitat of this species.

CLASSIFICATION

Geographic races: Robert Graves, an emeritus professor at Bowling Green State University in Ohio, devoted many years to the study of variation in this species. In his opinion, those in the southern half of Alberta are members of the *shelfordi* race, which he named to honour Victor Ernest Shelford, a tiger-beetle pioneer and distinguished ecologist. Those on the Lake Athabasca sand dunes in extreme northern Alberta and Saskatchewan are members of what Graves named the *athabascensis* race. Beach tigers in the intervening area (the Athabasca River drainage outside the Lake Athabasca dune fields) are difficult to classify, and Graves referred to these simply as "populations of uncertain status."

Colour morphs: Members of the *shelfordi* race are always brown, while about half of the *athabascensis* beetles are green or blue. Beach tigers in the Athabasca River drainage are usually brown, but occasional green or blue individuals occur here as well.

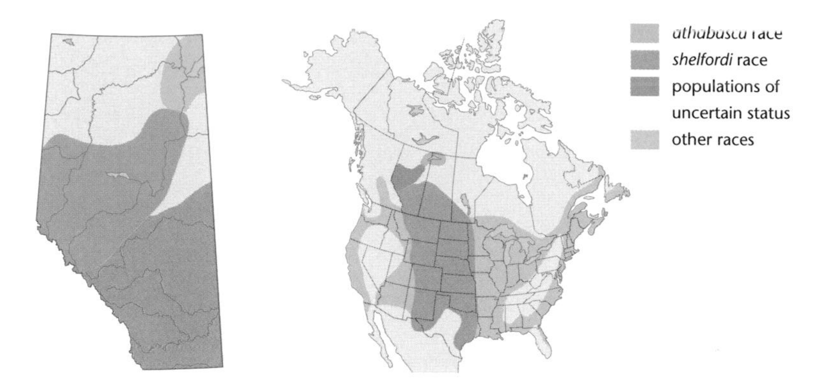

ECOLOGY

Habitat: These beetles live on dry, windblown sand, usually near water. Sand bars along big prairie rivers and beach dunes on the downwind side of lakes are both to their liking. On the Lake Athabasca sand dunes, where things are cooler, they roam farther from water than is typical elsewhere.

Life history: Adults emerge from the pupae in late summer, hibernate, and mate the following spring. Some tiger-beetle specialists have suggested that this species has non-hibernating summer adults, rather than overwintering ones, but this is not so to my knowledge. I was convinced of this when I found a female at Buffalo Lake on 23 September and took her home to a terrarium. When she died, I dissected her and found her ovaries undeveloped. She was clearly ready to hibernate, since the ovaries in overwintering beetles develop only after a cool period of "diapause."

Range: This species is widespread in Alberta, but localized.

Where to find them: Read the notes below, hope for the best, and then try the Buffalo Lake Recreation Area.

CICINDELOBILIA

I used to take this species for granted, despite the fact that my friends had trouble finding it. My family spent a few weeks each summer at Gull Lake, where the

Southern tiger-beetlers dream of seeing this northern rarity: the hyperborea *race of the sandy tiger.*

beach tiger was common. It seemed to prefer the dune ridge downwind of the plowed part of Aspen Beach. When it came time to photograph a beach tiger for this book, however, I ran into troubles. First, it became clear that they were no longer present at Gull Lake, since the dunes had grown over with willow bushes. So, I went up to Lac La Biche, where I had seen them on the beaches in the late 1970s. I suppose they might still have been there if there had been any beaches, but high waters in 1997 covered the habitat completely. Then I tried Lesser Slave Lake, where the habitat looked good but the beetles were mysteriously absent. Was I experiencing the summer doldrums, between generations of adult beetles? It seemed too early in the season for that. Re-reading Bob Graves' paper, I saw that he had been able to locate only three specimens from Lesser Slave Lake, so perhaps they are not common there. Farther south, a sand bar in the Red Deer River north of Bindloss was crawling with tiger beetles. However, only 1 out of 65 was a beach tiger, and it flew away before I could take its picture. I finally succeeded at the Buffalo Lake Recreation Area, where I found four of the beetles on two separate trips, and by that point I vowed never again to take these beetles for granted. On a similar note, despite its huge geographic range, this species is disappearing from many of its traditional haunts in the United States, largely due to beachfront development and habitat loss.

Sandy rhymes with randy! But why the males hold their front legs out to the side is still a mystery.

sandy tiger beetle

CICINDELA LIMBATA ("Siss-inn-DELL-ah lim-BATT-ah")

Nice sandy colour,
So unlike the other guys,
Good thing it's common.

IDENTIFICATION

In the field: Sandy tiger beetles from the southern two-thirds of Alberta have mainly white wing covers, a brown triangle down the middle of the back, one to three small brown splotches within the light part of the wing covers, and dark legs. In the north, the markings are more like those of an oblique tiger beetle, with a long shoulder mark and no distinct foot on the middle band.

In the hand: Hairy forehead, with underparts blue-green on the abdomen and coppery on the thorax.

Length: 10–12 mm.

Similar species: Southern sandy tigers look a bit like ghost tigers. The latter have a paler area down the middle of the back and pale legs, while the legs of the sandy are dark. Northern sandy tigers vaguely resemble oblique tigers, but the sandys are much smaller and have no "foot" to the middle band.

THE NAME

Limbata means edged or bordered, and presumably refers to the light parts of this beetle's wing covers. The name "sandy tiger beetle" refers both to the habitat of the species and the colour of typical southern beetles. The names of our two races make mythological references and mean "lady fairies of waters, meadows and forests" (*nympha*) and "beyond the North Wind" (*hyperborea*).

CLASSIFICATION

Geographic races: There are two distinct races in Alberta: the mostly white southern *nympha* race and the mostly brown northern *hyperborea* race. J.B. Wallis expected intermediate populations north of Edmonton, but I would expect them farther north, somewhere between Lac La Biche and Fort McMurray. John and Bert Carr discovered intermediate beetles near La Loche, Saskatchewan, on sandy road cuts that may have provided new habitat into which the two races dispersed from both north and south.

Colour morphs: In both races, one occasionally finds a greenish individual among the brown-marked masses.

ECOLOGY

Habitat: These beetles live on open sand, usually far from water. The sand does not have to be wind blown, but dunes and blowouts support the biggest populations.

Life history: Sandy tigers emerge in the first half of August, hibernate, and breed the following spring. In Manitoba, the larva takes two years to mature. Near Empress, in the dry summer of 1984, the adult sandys emerged in September, a full month later than the beautiful, blowout, and festive tigers.

Range: In the southern half of Alberta, but not as far west as the mountains, one sees the *nympha* race, while the *hyperborea* race is present from about the latitude of Fort McMurray north.

Where to find them: If you can find a roadside bit of open sand, you can probably find sandy tigers. Every day, thousands of people drive by good sandy-tiger habitat along Highway 2, in the Wolf Creek sand dunes north of Lacombe. The Opal Natural Area is another good spot (take the first left turn north of the town of Opal and continue until the road forks—take the left fork and park by the open sand). For the *hyperborea* race, the most accessible population lives in a sandy clearing in a pine and poplar forest about 30 km north of the bridge at Fort McMurray, up toward Fort McKay.

Female of the hyperborea *race, on sand south of Fort McKay.*

CICINDELOBILIA

When a new dune or road cut is formed, these are the first sand-dwelling tiger beetles to appear. They must fly around a lot, probably in late summer before overwintering, but so far we tiger-beetlers have yet to catch them in the act. If you notice a newly disturbed piece of sand, make note of the length of time it takes for this species to appear. The light colour of the southern race of this species seems to serve two functions. First, it matches the colour of the sand and makes the beetles more difficult to see. Second, it reflects sunlight and allows the beetles to forage longer on the hot sands. I did experiments to compare the amount of time that beetles from the two races spent foraging, basking, and cooling in the shade, and found that this was so. On the other hand, the northern race, with its darker pattern, warms up more readily in the sun and can forage longer in cooler places. It is noteworthy that the northern race's colour pattern is more like the ancestral pattern for the maritime group. The southern beetles are the more highly evolved ones ("derived" is the term biologists prefer). In the summer of 1984, on the Empress dunes in southern Alberta, it seemed that newly emerged beautiful tiger beetles were preying heavily on sandy tigers. The next summer I set about documenting this, but found no support for the idea at all. Either I was mistaken the first time, or it happens only under certain conditions. It is also noteworthy that tiger beetles do not become introduced species frequently. There is, however, an isolated population of the hyperborea *race at Goose Bay, Labrador, that most of us think arrived as a stowaway on a military airplane from northern Saskatchewan. This population is a bit greener than the norm, mind you, and at least one cicindelophile has argued that it is a naturally occurring relict population that constitutes a separate race.*

The Cowpath Group

THE FOLLOWING THREE SPECIES belong to another well-defined subgroup within the temperate tiger beetles. They take their name from the cowpath tiger beetle (*Cicindela purpurea*) and are typically colourful, appearing as adults relatively early in the spring and late in the fall.

Its head like a glittering jewel, a claybank tiger beetle displays its incredible colouration.

Now that's a good-looking beetle! A female claybank tiger from the river banks in Edmonton.

claybank tiger beetle

CICINDELA LIMBALIS ("Siss-inn-DELL-ah lim-BAH-liss")

Right there—a beetle,
Up close, white and purple-red,
That's why I get close!

IDENTIFICATION

In the field: The red head and PRONOTUM, vividly outlined in green, are good field marks for all colour morphs, as is the shoulder mark, which is reduced to two spots.

In the hand: Hairy forehead, underside of abdomen iridescent blue-green, undersides of head and pronotum bright red.

Length: 11–14 mm.

Similar species: Greenish individuals look a bit like badlands tigers, but are not as brilliant and do not have a green underside, head, and pronotum.

THE NAME

Limbalis means bordered or bordery, and is probably intended to mean the same as *limbata*. Presumably, the name refers to the light markings that border the wing covers, but it could also refer to the green edge of the wing covers. "Claybank" is a good way to remember both the beetle and its habitat.

CLASSIFICATION

Geographic races: In Alberta, all of our claybank tigers belong to the *limbalis* race.

Colour morphs: These beetles occur in two colour morphs here. One is bright red in ground colour, with green trim and crisp white markings. The other has dull greenish red wing covers, but is otherwise similar. In the older literature, the red ones were called the *awemeana* morph, while the green ones were called *spreta*. Since these names have no formal standing in the scientific scheme of things, and are difficult to remember (at least I think so), I prefer the terms bright and dull. *Awemeana,* by the way, was coined by Colonel Casey to honour Aweme, Manitoba (near Spruce Woods Provincial Park), where Norman Criddle made his pioneering observations of sand-dune tiger-beetle ecology. Little remains of Aweme itself, and even Percy Criddle, Norman's legendarily grumpy father, didn't know if it should sound like "Ah-weem," or "Ah-wee-mee." In his diaries, he complained that "whoever chose the name in question must be a pronounced idiot." Coincidentally, the word *spreta* means "scorned" or "despised," although John LeConte coined this name.

ECOLOGY

Habitat: These beetles are easiest to find on bare clay slopes. In this sense they are like badlands tigers, but claybank tigers seem to prefer places where the climate is a bit moister. Thus, in almost identical-looking places, claybank tigers are common in Edmonton, while badlands tigers are common in Drumheller. On occasion, single claybank tigers appear on flat areas with sparsely vegetated soils, but rarely on sand or salt flats.

Life history: Adults overwinter, but unlike the other members of the cowpath group, these beetles survive well into early summer in good numbers. In Manitoba, the larval stage takes two years; we can probably assume the same is true here.

Range: Widespread in Alberta, but not found on the prairies nor apparently in the eastern parklands. Probably does not occur in the far northeast—so far there are no records.

Where to find them: Sunlit banks of the North Saskatchewan River and its tributaries, in and around Edmonton, are as good a place to look as any. Over much of its range in Alberta, this is the most brightly coloured tiger beetle and is therefore well worth finding.

For a while, some people thought that this beetle was simply a geographic race of the splendid tiger beetle (C. splendida), *a species found in the American Midwest. Now we are back to recognizing the claybank as a distinct species. To my eye, this is an elegant beetle, although its colours are such that at a distance it looks just as brown as the oblique tigers with which it often shares its habitats. Up close, however, even the dull morph is resplendent, and I enjoy watching them clamber around on the uneven surfaces that they seem to prefer as hunting grounds. When you go looking for this species, be prepared to encounter such added excitements as red-sided garter snakes, large red-and-black jumping spiders in the genus* Phidippus, *and thick, gooey mud where springs have softened the clay soil.*

What a handsome couple! Two claybank tiger beetles share a meaningful moment on the mud.

But it's green! A female cowpath tiger beetle, in the species that was once called the "purple tiger beetle."

cowpath tiger beetle

CICINDELA PURPUREA ("Siss-inn-DELL-ah purr-purr-REE-ah")

Well that's a nice one,
How elegantly coloured,
But purple—No!

IDENTIFICATION

In the field: A medium-sized, all-black or all-green tiger beetle with small white markings and a white tip to the wing covers.

In the hand: Hairy forehead, underside of body black with a purple iridescence in the black morph; blue-green on the abdomen and coppery on the thorax in the green morph.

Length: 12–14 mm.

Similar species: Long-lipped and black-bellied tiger beetles resemble black cowpath tigers, but have dark tips to the wing covers and bald foreheads. Green cowpath tigers are more or less unmistakable due to their reduced white markings.

THE NAME

Since Albertan members of this species are either black or green, I simply couldn't go with the name "purple tiger beetle," despite its widespread usage elsewhere. The scientific name *purpurea* means "purple" of course, and the eastern *purpurea* race is, in fact, purple. I then thought about calling this species the "perplexing tiger beetle"—a pun on the word purple and a reference to how difficult it is to distinguish this species from the claybank tiger beetle in much of eastern North America. However, when the name cowpath appeared in the literature, I realized it was perfect for these beetles.

CLASSIFICATION

Geographic races: J.B. Wallis mistakenly thought that our prairie populations were intermediate between typical eastern dark beetles and western green ones. Our populations belong to the *auduboni* race.

Colour morphs: In Alberta, these beetles occur in two colour morphs, and both morphs are present in all populations, presumably interbreeding freely. One is black; the other is iridescent green.

ECOLOGY

Habitat: This prairie species is found wherever grassy vegetation is having a tough time completely covering the ground. There, beetles appear in small numbers, from time to time. Truly large populations occur, however, where a badlands valley meets the surrounding prairie, and sometimes in disturbed areas as well.

Life history: Adults of this species quickly become less common after the beginning of May, and are most abundant in late April and then again in September. They hibernate as adults and breed in the spring.

Range: This is a species of the prairies and the parklands in Alberta, but it is much more abundant on the prairies.

Where to find them: You will come across this species by accident sooner or later. If you want to go looking specifically for them, try the top of a badlands riverbank where it meets the prairie, perhaps in or near Drumheller.

CICINDELOBILIA

This is one of the few species of tiger beetles for which I have seen a published record of its mass. Gary Shook weighed one at 0.18 grams, which, when you think of it, is not very heavy. I am amazed at myself when I contemplate the amount of time I have spent in pursuit of creatures so tiny. When I have these thoughts, my only consolation is that people generally spend more money on, and get less enjoyment from, gemstones that are even smaller. *Karl Werner, author of the most*

A black morph cowpath tiger, showing the hairy forehead that distinguishes it from similar species.

expensive tiger beetle books I own, wrote that this species "rather delights in chilly weather." Now that's a bit of an overstatement, especially when you consider what "chilly weather" means here in Alberta. His two books on North American tiger beetles—the only complete treatment available at the present time—cost me about $600. Their value lies primarily in the photographs of pinned specimens, but many of these are mislabelled. Sadly, the text is short and sketchy. In part, these books motivated me to write one of my own, as an affordable alternative for local naturalists.

A lovely badlands tiger beetle with an unusually well-developed "foot" on its middle band.

badlands tiger beetle

CICINDELA DECEMNOTATA ("Siss-inn-DELL-ah DESS-em-no-TAY-tah")

That green, green beetle,
See the way it runs about?
Mad and happy too!

IDENTIFICATION

In the field: A bright, scintillating green beetle with relatively thick light markings, and often no clear distinction between the "foot" and the "lower leg" of the middle band.

In the hand: Hairy forehead, underparts entirely blue-green.

Length: 12–14 mm.

Similar species: Resembles the closely related claybank tiger, but that species is not all bright green, has red trim on the head and prothorax, and usually has a distinct "foot" on the middle band.

THE NAME

The name *decemnotata* means "ten-marked." I chose "badlands" to refer to the most obvious landscape feature in the places they are found.

CLASSIFICATION

Geographic races: There are no geographic races of the badlands tiger, and Alberta individuals all look much alike.

Colour morphs: The tiger-beetle literature reports that a violet-coloured morph is common in the Peace River district. Chris Fisher and I went in search of these exotic beauties, but all we found were typical all-green beetles. Looking through Wallis' book on the trip home, we read that badlands tigers from Whitehorse, Yukon, are also supposed to have a violaceous colouring to the wing covers. A few weeks later, John and Bert Carr showed me specimens from that area, but they were more brownish green, much like dull claybank tigers. Coincidentally, I photographed one of the beetles Chris and I found at Dunvegan, using a powerful flash, and to my surprise the beetle in the picture looked a bit purple—or perhaps more blue. Since then, I have seen bluish and purplish pinned specimens from the prairies. I wonder now if all badlands tigers have a blue-purple component to their iridescent colour, and if this is only evident in strong light or on some dead specimens. The only way to resolve the issue would be to travel back to the Peace Country, observe many living beetles under natural light, and leave the tinted sunglasses, the camera, and the killing jar behind!

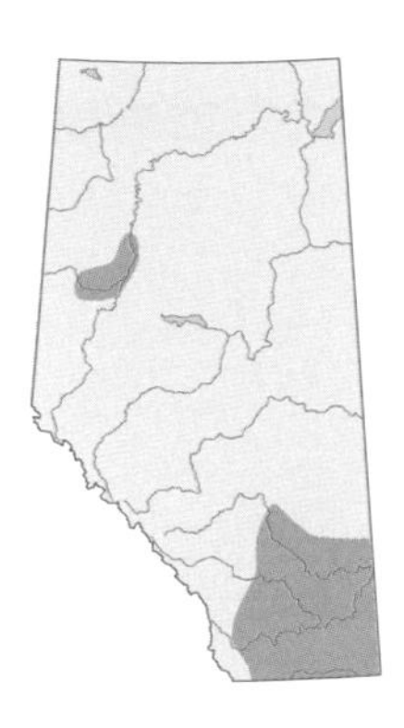

ECOLOGY

Habitat: Most common along the tops of prairie river valleys, on sparsely vegetated clay or gravel soil. In Alberta, the best sites are south facing, and the beetles are most common on brownish quaternary deposits (from the last ice age). These lie on top of the grey sandstone and clay that forms the dinosaur-rich badlands. On occasion, you will find badlands tigers in the valley bottoms on clay and gravel, or out on the bald-butt prairie, on open ground.

Life history: Adults emerge in the fall, later than most other tiger beetles, overwinter, and breed in the early spring, becoming noticeably less common by the end of May.

Range: In Alberta, this species lives in two regions: the prairies and the Peace River grasslands. Outside these two areas, in places where you might expect them, you only find the claybank tiger.

Where to find them: Walk along the north rim of any sparsely vegetated prairie river valley in late April or mid September and you should see these beetles. Access to many of these areas may require a bit of walking or climbing, but the scenery along the valley tops can be almost as breathtaking as the sight of the beetles themselves. The hillsides above

The mysterious "blue" beetle from near Dunvegan Provincial Park.

Drumheller, along the Dinosaur Trail, are a good place to look. In the Peace Country, try the big, bare hillside across the highway from Dunvegan Provincial Park.

CICINDELOBILIA

While not quite as breathtakingly beautiful as the festive tiger beetle, these are still among the finest insects in Alberta. Their splendour, combined with their appearance early and late in the season, when most insect-loving people are either not yet "into it" or mistakenly think the season is over, has given the badlands tiger something of a mystique, which it fully deserves. It took me a long time to find this species, since I expected it in Edmonton on the basis of an old and possibly faulty record in Wallis' book. This may have been based on a specimen in the University of Alberta collection that F.S. Carr collected in 1954. Two tiger-beetle experts have labelled it C. limbalis, *but it looks like an odd* C. decemnotata *to me. Perhaps it is the locality label that is in error. When I did find badlands tigers, down on the prairies, they seemed to show up one at a time, in places like the campground at Dinosaur Provincial Park and the edges of various prairie roads. Finally, I visited the right habitat at the right time of year, and they were as abundant as any other tiger beetles I have seen, alongside many cowpath tiger beetles in most places. Generally, the badlands and claybank tigers are not found on the same hillsides, but in the Peace River parklands their ranges overlap. In and around Dunvegan Provincial Park, I have seen the badlands tigers on the high, dry hillside, while the claybank tigers live lower down, closer to water. I have encountered only one of these beetles in an odd habitat, on an open sand dune, about a kilometre from a sizable population of badlands tigers, in the fall.*

The "Ungrouped Species"

WITHIN THE TEMPERATE TIGER BEETLE GROUP, there are a number of species that appear to have no close living relatives, or if they do, the related species do not live in Alberta. Thus I placed the following five species together by default and not because we have any evidence to suggest they are each other's closest relations. I should mention, however, that there have been suggestions to the contrary. Rivalier wrote that the beautiful and festive tigers (*C. formosa* and *C. scutellaris*) are related, and that the oblique and blowout tigers (*C. tranquebarica* and *C. lengi*) are as well. Some tiger-beetle experts also place the beautiful tiger in or near the cowpath group.

A freshly emerged festive tiger with a full head of hair that will soon wear off as it digs burrows.

Beautiful tigers live up to their name, with bright colours and the biggest bodies of all our species.

beautiful tiger beetle

CICINDELA FORMOSA ("Siss-inn-DELL-ah for-MO-sah")

Hot wind on the hills,
I recognize that buzz...
formosa *takes wing!*

IDENTIFICATION

In the field: A big tiger beetle (our biggest), with prominent white markings and bright-red ground colour.

In the hand: Hairy forehead and primarily blue iridescent underparts.

Length: 15–17 mm.

Similar species: At a distance, this species can look a lot like the blowout tiger beetle, but look for the shorter shoulder mark and thicker markings overall. Up close, size and underpart colour are obvious differences.

THE NAME

"Beautiful" is the English translation of *formosa*, and I couldn't agree more. The country of Taiwan used to be called Formosa (for the same reason), but this has nothing to do with the beetle.

CLASSIFICATION

Geographic races: Our Alberta populations have long been considered part of the *formosa* race. However, the white markings of this race cover about 30 to 40 percent of the wing covers, where ours measure in at closer to 50 percent. This marking makes them look an awful lot like the eastern race *manitoba*, except that they are more brownish where ours are red. Between Alberta and the range of the *manitoba* race, there lies the range of much lighter (65 to 95 percent white) *gibsoni* race, centred on the Great Sand Hills of Saskatchewan. So perhaps the Alberta populations are intergrades between the *formosa* race and the *gibsoni* race. Or it may be that the Alberta populations (like the Manitoba populations) have become whiter than their *formosa*-race ancestors in isolation and now constitute a race of their own. In any event, they are easy enough to recognize and I hope that we will soon have the answer to how they originated.

Colour morphs: There is only one colour morph of this species in Alberta, and the only obvious variation has to do with slight differences in the thickness of the white markings.

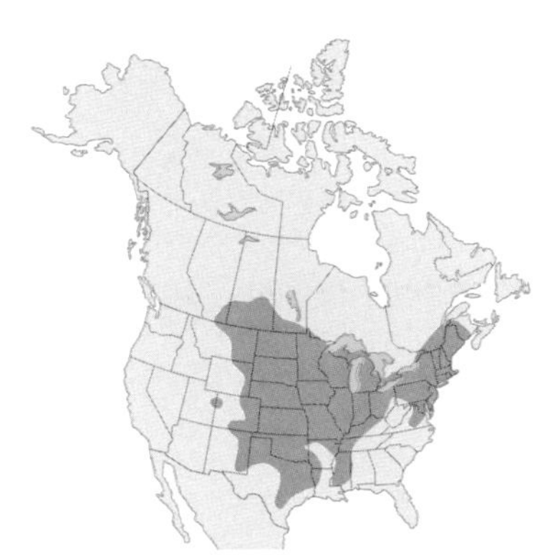

ECOLOGY

Habitat: Grant Gaumer wrote an excellent (although unpublished) PhD thesis on the beautiful tiger beetle for the Texas A & M University and characterized this species' habitat as deep, well-drained sand. In Alberta, they are prairie beetles that live along the margins of large dunes, as well as on blowouts. They are absent, however, from many sandy areas, such as the dunes at Pakowki Lake. It is not clear to me whether this reflects the beetles' inability to find the habitat or whether my idea of good habitat is different than theirs. Various naturalists have informally suggested that the ground water in this area is likely quite saline: perhaps this could explain the absence of beautiful tigers here.

Life history: Adults emerge in August, overwinter, and breed the following spring. In the US, the larval stage takes one year, but in Manitoba, Norman Criddle found that it takes two. It is likely that the Alberta populations are similar to those in Manitoba.

Range: Historically, this species has been found near Empress, Gem, Medicine Hat, Fort Macleod, and Sandy Point (where Highway 41 crosses the South Saskatchewan River). However, I know of no recent records outside the Empress region. Any such sightings would be welcome news indeed.

Where to find them: The most accessible location for this species in Alberta is a small sand blowout between Empress and the Empress cemetery. It lies along a dirt road that begins at the northeast end of town. The population here is likely small, so it would be wise for all of us to refrain from taking any of these beetles home.

This is a truly impressive beetle. It even has its own distinctive buzz when it flies. I found that the beautiful tigers are creatures of the scurf-pea zone, on the margin of sand dunes. They are capable of overpowering smaller tiger beetles, such as the sandy and the ghost, as well as digger wasps and other hefty prey that most other tiger beetles would find unmanageable. This is also, by the way, the only tiger beetle in Alberta that doesn't seem to have much trouble piercing the skin of a careless handler. 🐞 *I have mentioned the potentially unique nature of our Alberta populations. To get a sense of how isolated these beetles are, one has to travel around the sand hills in southeastern Alberta and southwestern Saskatchewan. North of Burstall (only 18 km from the Empress populations) there are lovely dunes, swarming with sandy, ghost, blowout and festive tigers, but strangely lacking in beautiful tigers. On two occasions, mind you, I saw but failed to catch what I thought were typical Albertan-looking individuals. Farther east, at the massive dunes south of Sceptre, Saskatchewan, the same is true—the party is on, but someone forgot to invite* formosa. *But when you get down to the dunes east of Fox Valley, even farther south, there are plenty of beautiful tigers all over the place—of the light-coloured* gibsoni *race, looking like giant purple-and-white sandy tigers. The distances separating these populations are small, and the only reasonable explanation for all of this complexity is that these beetles are not at all good at dispersing to new habitats. Instead, they are slowly doing their own evolutionary things on isolated sand-dune "islands" in a sea of grass and wheat.* 🐞 *This species supposedly possesses another distinguishing feature: its larval burrow, which is recessed in the side of an oblong pit. Other insects fall in the larva's pit, and what follows is like shooting fish in a barrel. All of our other tiger-beetle larvae dig burrows in which the entrance is flush with the ground surface. I must admit, however, that while I have seen this sort of burrow many times in Manitoba, I can't recall ever seeing it in Alberta. Perhaps our beautiful tiger larvae are more conservative in their habits.*

The pit at the entrance to a larval burrow in Manitoba. Do our Alberta beautiful tigers do the same?

Doesn't it remind you of Christmas? A well-coloured festive tiger beetle.

festive tiger beetle

CICINDELA SCUTELLARIS ("Siss-inn-DELL-ah SKOO-tell-AIR-iss")

There, in the scurf pea
Do you see the nice red one?
Oh my god it's bright!

IDENTIFICATION

In the field: A gorgeous, medium-sized, red-and-green tiger beetle with no light markings at all.

In the hand: Unmistakable, but note the sparse hairs on the forehead, and the entirely iridescent green-blue underside.

Length: 12–13 mm.

Similar species: None.

THE NAME

The word *scutellaris* means covered with small plates, and the SCUTELLUM is a small structure between the bases of a beetle's wing covers. I am not sure why the great nineteenth-century naturalist Thomas Say would have thought *scutellaris* an appropriate name for this beetle. I call it the "festive tiger beetle" because of its bright colours, and because red and green reminds me of Christmas. I once had a few in captivity that lived through the holiday season, and perhaps this also influenced my choice of the name.

CLASSIFICATION

Geographic races: There is only one race represented in Alberta, the *scutellaris* race. Elsewhere this is a highly variable species, but we needn't go into detail about that here.

Colour morphs: Most beetles are bright, iridescent red and green, but a few are noticeably duller. I think of these dull ones as a colour morph, but whether it is the result of genetics or something else, I have no idea.

ECOLOGY

Habitat: This is a species of sandy soils on the prairie, and it is most abundant where the only plant is the scurf pea, as on dune margins or the edges of large blowouts. It is also found in small numbers on sparsely vegetated sandy prairie and stabilized sand hills.

Life history: A typical temperate tiger beetle, the festive tiger emerges from the pupa in late summer, overwinters, and mates the following spring. The larvae take two years to develop in Manitoba, and probably here as well.

Range: This species appears to be widely distributed on sandy ground in the prairie region of Alberta, but it is likely only common in the Empress region. Records also exist for Medicine Hat (Police Point Park), Chappice Lake, Drumheller, Gem, the Suffield Military Reserve, and east of Bassano, but some of these reports may not represent viable populations at the present time. Still, it seems that this species is better able to colonize a variety of habitats than is the beautiful tiger.

Where to find them: The same Empress blowout that supports a small population of beautiful tigers is also home to the festive. However, it is also possible to see this species on the sand dunes directly south of Sceptre, Saskatchewan, in larger numbers and in more typical dune-margin habitat.

CICINDELOBILIA

It seems that the first thing anyone says when he or she become interested in tiger beetles is something like, "I really want to see those two: formosa *and* scutellaris!*" They are truly the great crowd-pleasers, and I must admit they are a hard pair to beat. When you see them hunting together on a dune margin, with the smell of sage in the air and the sound of Sprague's pipits overhead, life is good. Just remember to watch for rattlesnakes, and the experience should live up to all your*

The Nuttall's blister beetle, which I believe shares its warning colours with the festive tiger beetle.

expectations. *When I was analyzing the colour patterns of the dune tiger beetles in western Canada, this one really baffled me. Why no light markings? Why the bright colours? What caused its ancestors to leave the typical markings behind? Then I realized that while the festive tigers were zipping about under the scurf-pea plants on my favourite dunes, another bright red-and-green beetle was clumsily climbing the stems above. It was the blister beetle* Lytta nuttalli, *a toxic creature loaded with cantharidin. This is the same chemical that is found in Spanish Fly (the "fly" itself is a closely related blister beetle)—a so-called aphrodisiac that is actually a dangerous poison. So I proposed the notion that the tiger beetle and the blister beetle were mimics, sharing a warning colour pattern to let bird predators know they both taste bad. I was, at the time, quite pleased with myself for being the first to notice this obvious resemblance. As a beetle collector, I had filed my tiger-beetle experiences in one part of the brain and my blister-beetle experiences in another. It was very satisfying to suddenly start thinking like a beetle-eating bird and finally make the connection!*

Male shiny tiger up on tip-toes to keep its body high off the hot surface of the ground.

shiny tiger beetle

CICINDELA FULGIDA ("Siss-inn-DELL-ah FULL-jidd-ah")

Not really "greasy"
"Shining," nor "metallic," no...
Simply beetlish.

IDENTIFICATION

In the field: A small, polished-looking tiger beetle with bold white markings, including a long, thick shoulder mark that touches or nearly touches the middle band.

In the hand: Hairy forehead, iridescent green-blue underside.

Length: 10–12.5 mm.

Similar species: The blowout tiger is larger and less shiny than the shiny tiger, and is found on sand, not salt. Oblique tigers are much larger than shiny tigers, but since they occur in the same places, they may well be confused.

THE NAME

The word *fulgida* means shining, and since the name is so appropriate for this highly polished-looking beetle, it works in English as well.

CLASSIFICATION

Geographic races: Two very similar races supposedly meet in southern Alberta and Saskatchewan. In the Red Deer and South Saskatchewan drainages, our populations are like the *westbournei* race (named for a spot near

Westbourne, Manitoba). They are slightly smaller, more heavily marked, and less shiny than the southern *fulgida* race, known from the Milk River drainage. Harold Willis studied this species in detail, but had few specimens from Alberta, and so called all of the Saskatchewan populations intergrades between the two races. This creates the odd situation where "true" *westbournei* live only around Westbourne itself, while the intergrades range across much of southern Alberta and Saskatchewan, as well as northern Montana and North Dakota. Until someone sorts this all out, most tiger-beetle people have quietly decided to call the Milk River beetles part of the *fulgida* race and those from farther north *westbournei*, even if this distorts the original, intended meaning of the *westbournei* name. Here, I choose to avoid the use of subspecies names.

Colour morphs: Most of the shiny tiger beetles in Alberta are reddish green or greenish red, but from time to time we find blue, bright-green, and even purple specimens.

ECOLOGY

Habitat: This beetle lives on bare, salty places ("alkali flats," despite the fact that they are not necessarily alkaline), but not just any bare, salty place will do. They occur most abundantly where moisture is flowing through the soil, from a nearby creek, river, or spring-fed seep.

Life history: This is another spring-fall species, despite the fact that it is almost always found alongside summer-adult species, the salt creek and grass-runner tigers. The larval lifespan is unknown in Alberta, but is probably two years.

Range: These beetles are restricted to the prairie region here in Alberta and are most abundant along the valleys of the major rivers.

Where to find them: Try the Jenner Bridge site (described under the salt creek tiger) or the white, salty areas just upslope of the Sandy Point campground, where Highway 41 crosses the South Saskatchewan River. Watch among the sparse plants and be prepared to get your feet dirty if the ground is wet. This spot used to be a biker hangout and could get rowdy at times, but has been much more sedate since the campground was upgraded. I didn't realize it was such a good location for tiger beetles until I was driving over the bridge and spotted John and Bert Carr with their nets. They were chasing beetles with Chris and Ann Van Nidek, tiger-beetle enthusiasts visiting from the Netherlands. That day we went down to the Great Sand Hills of Saskatchewan and found the *gibsoni* race of the beautiful tiger. We dined at Lynda's Pizza in Burstall, Saskatchewan—a place every tiger-beetler eventually comes to hold dear.

CICINDELOBILIA

When J.B. Wallis' book first appeared, this species was known only from one locality in Alberta, near Onefour along the Lost River. George Ball and the legendary ornithologist Bob Lister found it there in the late 1950s. When I first became interested in tiger beetles, my friends and I all assumed that the shiny tiger was one of those extremely rare southeastern creatures, like short-horned lizards and painted turtles. I now think the beetle is relatively abundant and widely distributed in southeastern Alberta. This is not to say we should take it for granted, mind you. In British Columbia, its close relative, Cicindela parowana, *has apparently disappeared. Its Okanagan salt flats have been bulldozed and planted to grapes.* ✱ *Harold L. Willis, in a classic study of the tiger beetles of saline habitats in the central United States, remarked that this species is typical of cool, northern climes. I guess that may be true among the salt-loving sorts of tiger beetles, but to me this is a southern creature that lives only in the warmest and driest parts of Alberta. Mind you, Willis also cites reports of salt-flat temperatures as high as 80°C, a full 23°C hotter than any surface temperature I have ever recorded in western Canada. For the record, the warmest salt flat temperature I have recorded in Alberta was 38°C, at Sandy Point—on a typical summer afternoon in the cool domain of the so-called "northern" shiny tiger beetle.*

Trust me, it's the green morph. Lighting can make a tiger beetle's colours change with the angle.

The blowout tiger beetle in some sparse grass on clean white sand at the Opal Natural Area.

blowout tiger beetle

CICINDELA LENGI ("Siss-inn-DELL-ah LENG-eye")

Warm sand, smells of pine
We walked for miles, just talking
Did you see lengi?

IDENTIFICATION

In the field: A handsome, medium-large tiger beetle, usually red in ground colour, with light, well-developed markings and a long, oblique shoulder mark that is thickest near the shoulder itself.

In the hand: Hairy forehead, elongate upper lip (although not as long as in the sylvan group), underside of body iridescent copper, blue, purple, and green.

Length: 12–14 mm.

Similar species: Four Alberta species share similar markings. Blowout tigers are most like beautiful tigers, but they are larger, and in Alberta they have stubbier shoulder marks and blue undersides. Shiny tigers are smaller and shinier, and live on salt flats, not open sand. Oblique tigers have a shorter upper lip and a shoulder mark that is narrow near the shoulder itself. The *hyperborea* race of the sandy tiger is smaller, rounder, and only vaguely reminiscent of even the darkest of blowout tigers.

THE NAME

Charles William Leng (1859-1941) was Director of the Staten Island Institute of Arts and Sciences. In 1902, he wrote *Revision of Cicindelidae of Boreal America*. Much later, André Larochelle, a Québecois tiger-beetler, wrote, "He was a genial man, always quick to render aid to anyone, and he remains an inspiration and a model for all who are interested in the study of tiger beetles." The only photo I have seen of Leng shows him with his net in a rocky field, a small-town church in the background. With dress shirt, tie, and flat-brimmed hat, he looks like a tall, slim version of W.C. Fields. Such is the stuff of tiger-beetle legend. I chose the name "blowout tiger beetle" in recognition of this species' willingness to live on even the smallest openings in otherwise vegetated sand hills.

CLASSIFICATION

Geographic races: All of the Alberta blowout tigers belong to the widespread *versuta* race. In Manitoba, members of the same race are greenish on the head and pronotum, not red, and are duller than Alberta beetles. The same is true of Manitoba's beautiful tigers. The beetles' colours may match their ancestral soils: redder-in-the-neck here in Alberta and less flashy in Manitoba. Just before this was written, however, I reared two blowout tigers from Opal, Alberta, in a terrarium in my basement. The adults that emerged looked for all the world as if they came from good old Aweme, Manitoba. Presumably, something about the conditions during pupation influenced the colour of the beetles.

Colour morphs: In most places, most of these beetles will be reddish, but in some spots one can find blue, green, or even black individuals (though still with the typical white markings). In the Rolling Hills, the blue morph is common, and Gerry Hilchie tells me it is found regularly near Chappice Lake as well (north of Medicine Hat on Highway 41). At Opal, I have seen a number of black individuals, some with very thin middle bands. John and Bert Carr tell me that the black morph becomes much more common farther north.

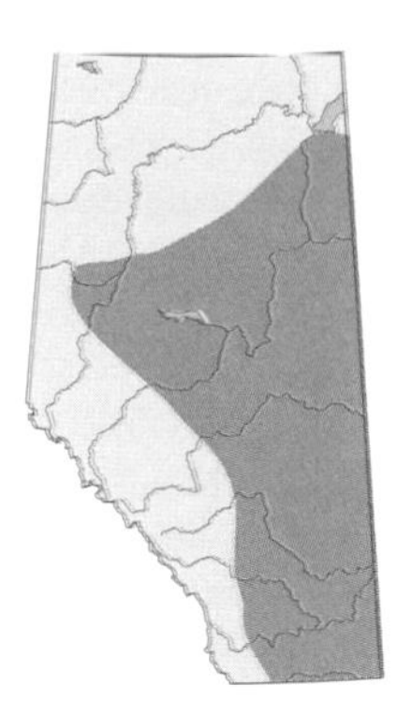

ECOLOGY

Habitat: These beetles live on dry sand, away from water. In the prairies, they live on the open dunes (where they are most common in the dune

Blue morph blowout tigers face to face in a non-violent moment that lasted only a fraction of a second.

margins, among scurf-pea plants), as well as roadside blowouts that support few other tiger beetles. In the parklands and boreal forest, they frequent open sandy places among jack pines, in the company of the long-lipped tiger beetle.

Life history: Blowout tiger beetles appear as adults in late summer and overwinter, reproducing the following spring and living into June or July. They probably spend two years as a larva, based on Norman Criddle's work in Manitoba.

Range: Records exist for almost all sandy areas in the eastern half of the province. It is likely that they can be found on any open sand area outside the mountains.

Where to find them: In southern Alberta, blowouts in the Hilda sand hills are a good bet (along Highway 41, about five kilometres north of the town of Hilda). You could also go to the sand dunes south of Sceptre, Saskatchewan, where the species is hugely abundant. In northern Alberta, try the Opal Natural Area, an unmarked parcel of crown land that is easily accessible by taking the first left turn north of the town of Opal and continuing until the road forks—take the left fork and park by the open sand.

Blowout tigers are good-sized, easy to find, and a popular species with tiger-beetlers, since they sport a bit of colour. Mind you, even a bright one isn't as brilliantly iridescent as a tiger beetle can be. Still, on the sands of a jack pine hill the red blowout tigers are a nice complement to the black long-lipped and the pale sandy. However, I don't think you can beat the sight of blue-morph beetles on the blowouts along the often rutted and mucky road (at other times smooth and dusty) that leads south from Lake Newell into the Rolling Hills. ✱ *While searching the sand hills of the prairies, I have come time and time again to places where former dunes lay covered by plants. These places sadden me. I am always cheering for the west wind in its continual struggle with the roots of the scurf pea, the sand grass, and the skeleton weed. When the sand loses the contest, so do the tiger beetles, but the last one to throw in the towel is the blowout tiger. On many a beetling trip I came away with nothing but a few memories of this stalwart animal, waiting patiently for drought, fire, or bulldozers to bring back the good old days once more.*

The rare black or dark morph of the blowout tiger, with its typically scrawny middle band.

Too bad they aren't colourful, because oblique tigers are big, snappily marked, and omnipresent.

oblique tiger beetle

CICINDELA TRANQUEBARICA ("Siss-inn-DELL-ah TRAN-kweh-BARR-ick-ah")

Nighttime—flip the stone,
Flashlight illuminates the scene,
And a sleeping "Trank."

IDENTIFICATION

In the field: A dark, medium-large tiger beetle with well-developed markings, including a long, oblique shoulder mark.

In the hand: Hairy forehead, short upper lip; iridescent underparts red-purple on the head and prothorax, deep blue-green on the abdomen.

Length: 13–15 mm.

Similar species: Shiny tigers are smaller and much more polished looking. Blowout tigers are usually reddish and their shoulder marks are not narrowed at the shoulder itself. The *hyperborea* race of the sandy tiger is a bit like a smaller, flatter, more rounded oblique tiger. It has much thicker light markings and a more iridescent head and pronotum.

THE NAME

Oddly enough, this species was named for the town of Tranquebar, India. Perhaps a mislabelled specimen was to blame, at a time when many of the experts kept busy naming specimens brought to them by others. I grew up calling these beetles "tranks," but I think the name "oblique" is a better reminder of its most characteristic feature. I didn't want to call it anything like "common" or "ubiquitous," since these terms are invariably construed to mean "uninteresting" or "everyday." Luckily, of the names formerly given to this species, the worst, *Cicindela vulgaris* (which means common and usual), is now considered invalid. I wouldn't mind, on the other hand, if some taxonomist were to magically resurrect the name *obliquata*, which makes so much more sense (assuming it means "slanting sideways") than *tranquebarica*.

CLASSIFICATION

Geographic races: In general, one finds beetles with thinner light markings on the wing covers as one goes farther north. The northern beetles have been called the *borealis* race, while the southern ones have been called *kirbyi*. However, most tiger-beetle specialists would rather wait for someone to do a good, thorough study of the matter before they start using these names with much confidence.

Colour morphs: Most oblique tiger beetles are dark blackish brown, but on occasion one does run into a specimen that has a bit of an olive-green sheen to it or is just plain green. Greenishness shows up in many sorts of tiger beetles at the edges of the species' range, typically in high altitudes or northern locations. This makes many tiger-beetlers suspect that the environment, not genetics, causes some types of green color.

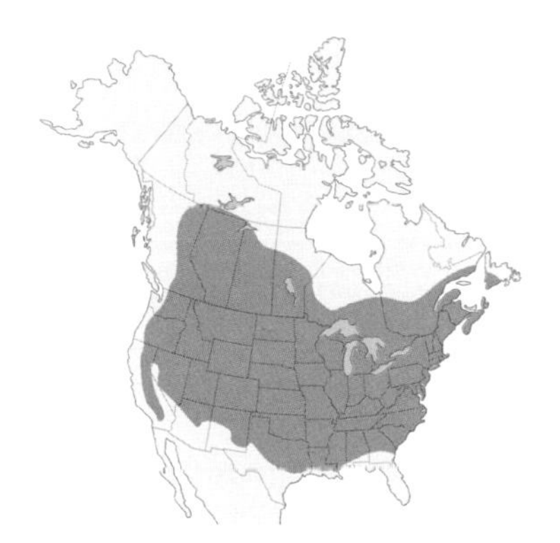

ECOLOGY

Habitat: Almost any place that is home to tiger beetles is home to obliques. I have found them on mudflats, claybanks, sand dunes, blowouts, salt flats, sparse prairies, and vacant lots. Still, the largest concentrations I have encountered have been on sandy clay soil on south-facing riverbanks, so perhaps they do have a habitat preference after all. Ecologists don't like the word "preference," by the way, since it implies something about the motives inside a tiger beetle's wee brain; they prefer "habitat associations" instead.

Life history: This is another spring-fall species that is sometimes hard to find in midsummer before the new adults emerge. As with all other Alberta species, we do not know the length of the larval period here, but it is probably about two years.

Range: Found throughout Alberta.

An oblique tiger beetle on salty clay: one of its many habitats in Alberta and elsewhere.

Where to find them: If you can find tiger beetles at all, you should be able to find this species. If you are searching specifically for the oblique tiger, I would suggest any bare, south-facing riverbank, upslope from any mud or sand flats. This is probably our most common and widespread tiger beetle.

CICINDELOBILIA

Since it is such a widespread, abundant species, the oblique tiger beetle is better studied than most of the other species in North America. Here in Alberta, for example, Janice Kuster compared this species—a typical daytime tiger beetle—to the ghost tiger, that is active after dark as well as by day. She found that the eyes of the oblique tiger are optimized for spotting prey in bright sunshine. In 1985, K.R. Morgan published an interesting study on the means by which these beetles regulate their body temperature in order to remain as active as possible without overheating. This study inspired many more along the same lines, including my own, although I don't believe anyone else re-focussed on this particular species. Still, many aspects of this beetle's biology remain a mystery, including the delineation of our local races. We do tend to take these beetles for granted. Instead of simply labelling them as "widespread and common," perhaps we should be asking how they cope in places where all others fail. Are obliques less fussy about their habitat requirements, or are they somehow more adaptable in the face of conditions that other species cannot tolerate?

A pinned specimen of Playchile pallida, *an oddball tiger beetle from southern Africa.*

4 A Few Reassuring Words About Tiger-Beetle Classification

LET ME WARN YOU: not long after you become interested in tiger beetles, someone will inform you that they are not a real family of beetles at all. "They are just ground beetles," you will be told. For some tiger-beetlers, the suggestion that their favourite creatures are nothing more than ground beetles has not gone over very well, but I can assure you this is not something to worry about.

All tiger beetles are predatory, and they all possess a unique type of jaws as adults. Their LARVAE (the young tiger-beetle "grubs") are unlike anything else in the beetle world and are remarkably similar throughout the group. Mind you, not all tiger beetles look like our Albertan varieties. The most primitive-looking tiger beetles are black, small-eyed creatures that are active at night. The next time you are on the British Columbia coast in early summer, you might check under boards at the edges of forests for *Omus*, one of the most primitive tiger beetles of all. At the other extreme, some tropical tiger beetles possess huge eyes, domed faces, and ant-like bodies, and live in trees.

It is clear that the tiger beetles belong in the predatory subgroup of beetles called the Adephaga, along with such things as ground beetles, whirligig beetles, diving beetles, and a number of other less-well-known groups. Modern systematic studies (the reconstruction of a group's evolutionary tree and the creation of a classification to go with it) have placed the tiger beetles just about everywhere imaginable within or near the family tree of ground beetles.

To determine whether they should be called a family or a subfamily, we need to know whether the ancestor of the first tiger beetle was a ground beetle (making tigers a subgroup of the ground beetles) or a more primitive member of the Adephaga (making them a group on their own). As of right now, the correct placement of the tiger beetles is more or less up for grabs.

My own suspicion, which is shared by many but by no means most other beetle scientists, is that the tiger beetles are distinctive enough and

Above left: John Acorn, Bert and John Carr, Anna and Chris Van Nidek in the Great Sand Hills of Saskatchewan in 1986. Above right: John Acorn, Bill Barr, Felix Sperling, Ted Pike, Kate Shaw, Janice Kuster, and Gerry Hilchie, on a 1976 expedition.

widespread enough likely to have been an early side branch of the adephagan family tree, before the evolution of ground beetles. For that reason I continue to think of them as a full-fledged family. But, whether you refer to them as the family Cicindelidae ("cicindelids" in colloquial English), or the subfamily Cicindelinae ("cicindelines"), they are still tiger beetles and they still evolved from something that was not. It doesn't make them any less tiger beetlish, or any less interesting, and it also doesn't hurt to learn a bit more about other beetles as you pursue an interest in the tigers.

Another potential source of confusion involves the fact that all of our Alberta tiger beetles were originally placed in the single huge genus *Cicindela*. In my opinion, this is a reasonable arrangement, since the species of *Cicindela* are all very much alike in both form and habits. It is not surprising, however, that entomologists have looked for subgroups within the genus, in order to place closely related species together.

The origin of these additional group names lies with the work of the French entomologist E. Rivalier, who looked at the male genitalia of tiger beetles and based his classification on resemblances among these. Rivalier's subdivisions of the genus have remained more or less unchallenged since the 1950s, with some authors treating them as full genera and others calling them subgenera. As well, even finer groupings of species within the subgenus *Cicindela* have been identified, primarily by Rick Freitag and his students at Lakehead University in Ontario.

All of these groupings serve only to make it easier for people interested in tiger beetles to talk about them using names that refer to groups of beetles that constitute whole branches (of whatever size) of the tiger beetle family tree. For the moment, the majority opinion has it that all of our species can be called *Cicindela*, and that the tiger beetles form a family of their own. If this changes, don't let it trouble you. After all, the beetles will stay the same.

Other fine people who have contributed to our knowledge of Alberta tiger beetles (clockwise from top left): George Ball, Don Shaw, Tim Spanton, and Rick Freitag.

Su-Ling Goh, biologist turned TV reporter, spends quality time on a muddy hillside, tiger-beetling.

5 The Joys Of Tiger-Beetling

TIGER BEETLES IN THE FIELD

As beetles go, tiger beetles are big and beautiful. They are also among the wariest of all beetles, so their splendour goes unnoticed by most people. The simplest solution to this dilemma is to catch them with a net.

Now this may sound easy, but the first rule of tiger-beetling is never, ever underestimate the sneakiness of a tiger beetle. Start with a good net (see the appendices for sources of equipment), and find yourself a beetle. They usually allow you to approach within about a body length without flying away. Your body length, that is, not theirs. To get closer, move in slow, smooth, Tai Chi-like fashion, crouching lower and lower as you approach. Then pounce, with all the speed you can muster.

Once the beetle is under the net, you have to find it very quickly and grasp it through the net fibre. Don't worry about being bitten—it's no big deal. The beetle may well regurgitate the contents of its crop at this point, and a well-seasoned net will have many small brown stains as memorabilia of fine catches. Once the beetle is secure, reach in with the other hand, grasp the beetle, and remove it from the net.

Tiger beetles are easy to handle, as long as you don't try to hold them by one leg or by the antennae. Then the appendages break off. Almost any other grip will work, and they can be held by the head, the sides of the body, the two hind legs, or any combination thereof. Once in the hand, the beetles can be examined with a magnifying glass or one side of your binoculars, turned backward. I like to carry a good quality 10-power ("10X") hand lens (the best are made with what is called "triplet" construction): with one of these you can see all of the features mentioned in this book and then some. Just remember to hold the lens to your eye first, then move the beetle in close with the other hand.

I have caught and released thousands (maybe even hundreds of thousands) of tiger beetles this way, and I know of no evidence that it harms them or reduces their chances of survival. In fact, I always let them bite me on the fingertip, just so they can get back at me for the inconvenience I have caused.

Of course, not everyone releases the beetles they have caught. The traditional way to enjoy tiger beetles is to make a collection of dead, dried, pinned specimens, and this is a legitimate thing to do if your motives are scientific and you plan to make good use of each and every specimen.

Otherwise, I see no reason not to identify the beetles in the field and let them go. Perhaps you will find that a small, synoptic collection (with one or two of each species or race) is a useful reference tool. I certainly found my own collection indispensable while writing this book. In any event, the techniques for making a beetle collection are widely publicized and need not be repeated here. I should also mention that while collecting our local tiger beetles is no longer necessary for identification, this is not yet the case in many parts of the world.

Of course, you can also watch tiger beetles the way birders watch birds. Binoculars will give you a much better view of a wary tiger beetle, despite the beetle's small size. Some binoculars focus quite close (less than two metres away), while others cannot focus much closer than four metres. Obviously, closer-focussing binoculars are better for tiger beetles. When I am in the field, I often take my Bausch and Lomb 10 X 42 Elite Waterproof binoculars. They focus so closely I can examine my own toenails! However, at the closest focussing distance, the visual fields of the two sides of the binocular do not overlap very much—in other words, you only see the beetle (or the toenail) with one eye at a time.

The answer to this problem is to start with a pair of compact binoculars of the "reverse-porro-prism" design. Their front lenses are closer together than the eyepiece lenses. If you then purchase a Nikon 5T close-up lens (a "two-element" lens—very important), you can put it in front of the binoculars and create a close-focussing binocular with excellent stereo vision.

Proper form, low to the ground, just before the pounce.
photo: Chris Fisher

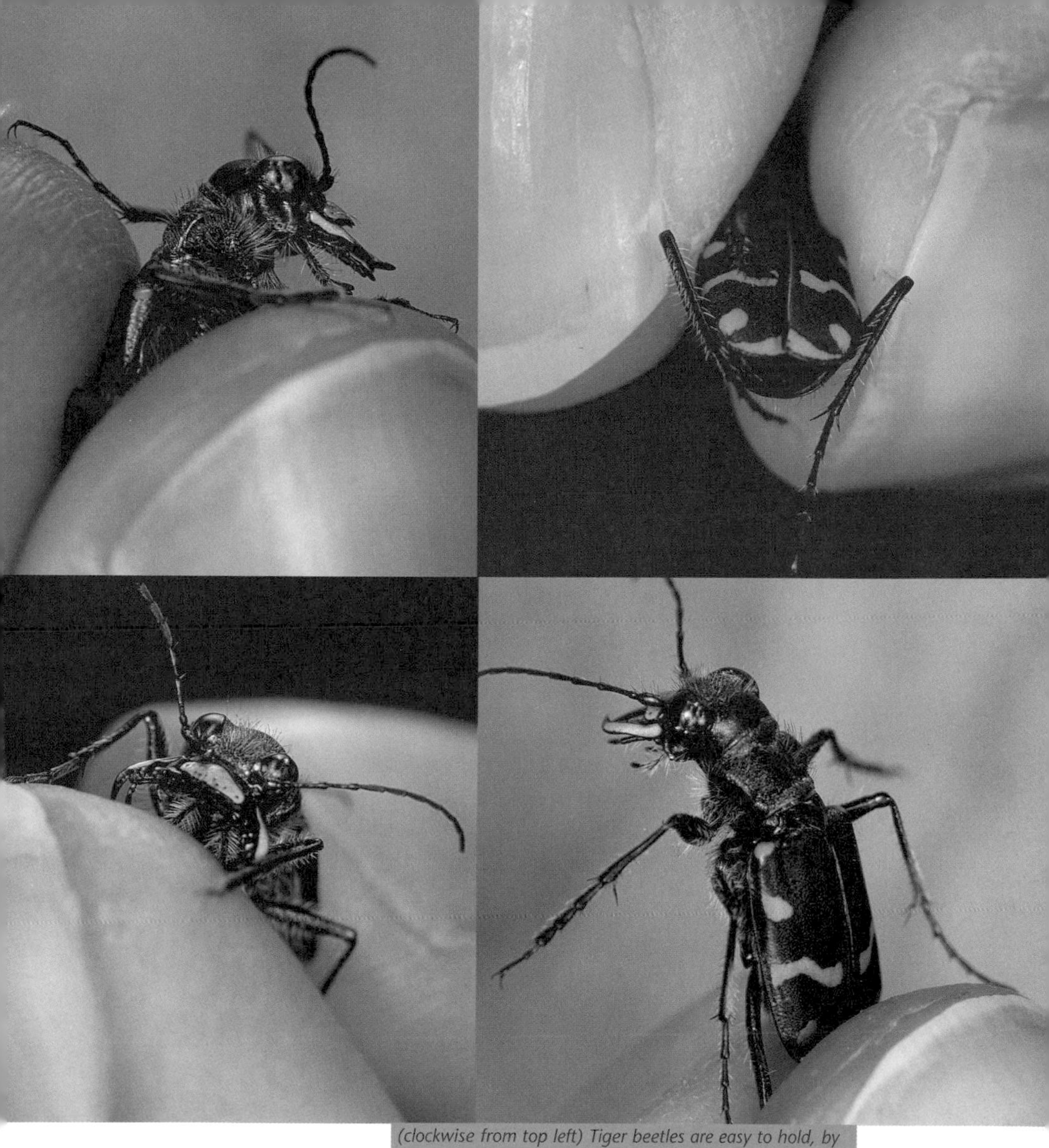

(clockwise from top left) Tiger beetles are easy to hold, by the body, the head, or the two hind legs. Don't worry, most of them can't pierce your skin (with thanks to hand model Carole Patterson).

I simply mount the close-up lens on a rubber lens hood (for use on a camera), and attach the lens hood to the binoculars with a couple of elastic bands. This whole arrangement, which was suggested to me by my friend Carroll Perkins, is perfect for examining insects of any sort, at a distance of about one-third to one-half a metre.

(top): Side (left) and front (right) view of the "Carroll Perkins Gizmo," a homemade close-focussing binocular. (bottom): Close-focussing monoculars are great, although you use them with only one eye at a time.

I should also mention the use of close-focussing monoculars, of which there are a few models on the market. These focus even closer than the Carroll Perkins gizmo, but they obviously cannot be used with more than one eye at a time. I like the Bushnell model 14-8200, an 8 X 20 model which I bought for about $45. They apparently sell quite a few to parking-meter attendants, who use them to check meters at a distance, but for us tiger-beetlers, they are perfect, cheap, and small enough to carry anywhere.

TIGER-BEETLE PHOTOGRAPHY

Tiger-beetle photography offers its own set of challenges, and it is also a fine alternative to collecting. To get started, you should definitely have a single lens reflex ("SLR") camera, so that what you see through the viewfinder is the same as what will appear on the film. There are many models available. I now use expensive auto-focus Nikon cameras, but it is possible to achieve similar results with "old-fashioned" manual-focus equipment. Used manual-focus cameras and lenses are inexpensive and often carry a limited warranty.

You need high magnification, and you can get it in a number of ways. You can buy a "macro" or "micro" lens, although some do not provide enough magnification for things the size of tiger beetles. The idea is to achieve a magnification of between 1:1 and 2:1, where the beetle is between life-size and half-size on the film. You can also modify an existing lens with either screw-on close-up lenses or extension tubes that fit between the camera and the primary lens. Either method will do the trick. Teleconverters will also increase the magnification of a lens.

Once you have achieved the right degree of magnification, you need to consider depth of field: how much of the beetle will be in focus. To get acceptable results, you will probably have to set your lens aperture to at least f16 (since a smaller lens opening gives more depth of field). I shoot almost all my macro photographs at f16. At this point, the lens is not letting in much light, so you may find you also need added illumination from an electronic flash. If you use a single flash, the effect should look a lot like natural sunlight.

The only other important consideration is working distance, or the distance between the camera and the beetle. With a creature as wary as a tiger beetle, getting in close enough for a good picture is not easy. I like a working

My "big gun" photo rig, with a 200-mm micro lens, a 2X teleconverter, and two flashes. photo: Chris Fisher

Photographers converging on a cooled tiger beetle, awaiting a natural pose.

distance of at least half a metre, so I use a 200-mm micro lens, a 2X teleconverter, and two flashes, mounted at close to the same axis. I need two flashes, since one does not throw enough light for this set-up. If you use a shorter lens, say a 50 mm with extension tubes, you will almost certainly have to cool your beetles to photograph them, or shoot them in a glass terrarium. (By cooling, I mean placing the beetles in a refrigerator (not the freezer!) for about five to ten minutes, after which they will be a bit slower than usual and somewhat easier to follow with a camera.) Whatever technique you choose, you should find tiger beetles superb subjects for nature photography. The purists will tell you that only a wild tiger beetle, photographed in the field, is a "legitimate" subject. As far as I'm concerned, however, if you start by getting any decent results at all, you can build on your skills and eventually find yourself able to photograph the beetles under whatever conditions you might encounter.

KEEPING TIGER BEETLES IN TERRARIA

There is only one way to know tiger beetles and that is to watch them in the field. See, feel, and smell the places they call home, and realize that they sense these things too—somewhat differently, of course, but heat is still heat, and wind is still wind. Accurate ideas about tiger-beetle lives are much easier to come by on a sunny afternoon by the river than in an armchair, working from memory.

There are, however, times when it is fun to have a few live tigers around the house. They do well in terraria, and their requirements are simple. The main thing they need is drinking water. A few days without and most will die for sure.

The Shaw family once kept a few sandy tiger beetles alive for months in a narrow-mouthed terrarium jar half full of potting soil, on an end table in the living room. This demonstrated to me that tiger beetles don't need much light, much warmth, or the same type of soil they live on in nature. Just because they can survive in odd cages, however, doesn't mean this is the best way to keep them. I put mine in the smallest size of all-glass aquarium available, sold as "two-and-half-gallon" tanks but really much smaller. I give them about eight centimetres of clean sand or clay to run on (their preferred soil in nature), and I keep one corner of the tank wet so they can always drink. That means wetting the corner once a day, every day. I buy a light hood for each tank, and in it I place a 25 or 40-watt aquarium bulb. The hood provides enough heat and light for the tiger beetles to behave normally and also prevents them from escaping. With a light overhead, they seem content to stay in the tank—lit from one side only, they spend their time trying to reach the light. For food, I give the beetles insects, worms, crickets, or dry dog food soaked in warm water. One feeding every two days seems sufficient.

With this method, the beetles live as long in captivity as they would in the wild. For example, I caught three festive tigers in late August last year, and kept them alive for observation and photography. In the wild, they would have stayed out for another four to six weeks, hibernated, and emerged

A simple, elegant tiger-beetle terrarium: an understated item that will harmonize with almost any decor.

A thirsty bronzed tiger sipping water from wet sand in one corner of its terrarium.

in early April, ready to live until some time in June or July. Not counting hibernation, they could look forward to three or four more months of active life. In the terrarium, they were looking great at Christmas, and the last one died in February, well into its golden age.

I didn't allow those beetles to hibernate, but inducing hibernation is simple. Starve them for a week so food won't go bad in their guts, reduce the day length to 10 to 12 hours, and then put the entire terrarium in a cool attached garage or the refrigerator for six weeks. When they emerge, give them 16-hour days, and they will soon mate.

You can see interesting things when you keep tiger beetles under close observation. For example, festive tigers spend a lot of time hiding underground. Male salt creek tigers spend almost all day riding around on the females, guarding them from other males.

If you are really lucky, your beetles will eventually give you some larvae. You probably won't see the eggs, but one day there will be little dark heads at the tops of circular burrows all over the soil surface. Larvae are less prone to dehydration than adults, and more than once I have dumped out a jar of bone-dry sand only to uncover a hungry larva.

Place each larva in a tall jar with appropriate soil, and give it insects to eat. Fruit flies work for the small larvae; try crickets for the bigger ones. Larvae will also take small bits of raw hamburger or dog food placed gently in the entrance of their burrow. Keep the soil moist but not wet, and provide both light and heat, as with the adults. You should be able to accelerate the larval development from two years to eight months or so. Remember that they close their burrows from time to time—when a larva disappears it does not mean that it has died. It may simply be waiting for you to improve conditions in the rearing jar.

I doubt that keeping tiger beetles will ever be as popular as fish- or reptile-keeping (woe to the beetles if it does), but observing them in terraria does have its delights. In general, however, tiger beetles are too short lived and require too much daily care to appeal to any but the most devoted terrarium keepers. But if you come to love tiger beetles, you'll want to try it, and if you try it, you'll want to do it right.

Mandible marks in wet mud, where captive salt creek tiger beetles were drinking.

Rare habitat: a prairie sand dune along the valley of the South Saskatchewan River.

6

Should We Be Protecting the Tiger Beetles of Alberta?

HERE IN NORTH AMERICA, it would seem that tiger beetles are among our most endangered insects. Conservationists have sounded the alarm on behalf of approximately 30 species and races in the United States and Canada alone. Surely they must be suffering badly at the hands of humanity.

Sadly, I think this is indeed the case, at least in localized areas. Changes in drainage patterns reduce the amount of shoreline habitat available to tigers, erosion control reduces the amount of sand dune and blowout habitat, and both agriculture and urban sprawl have a way of replacing all manner of tiger-beetle habitats with various sorts of familiar, lifeless landscapes. Admittedly, some types of construction create habitat for these beetles, such as open road cuts, sand and gravel pits, bare vacant lots, and the bottoms of drained water bodies. Still, the opinion of tiger-beetle specialists seems to be that we are losing beetles and their habitats faster than we are gaining them.

Beetle collecting, by the way, has never been shown to significantly affect tiger-beetle populations. In theory it could do so if enough people wanted a particular sort of beetle badly, and they all converged on a critical population in a small isolated area, just as the adults were emerging from the pupae. As far as I know, however, this has yet to happen, despite the fact that some not-so-conscientious collectors have collected extremely heavily from a number of small, isolated tiger-beetle populations, particularly in the United States.

So, what about Alberta's beetles? After much thought, the only tigers I have concerns for in the province are the ghost tiger beetle (see page 28) and the beautiful tiger beetle (see page 66) (assuming that the *imperfecta* race of the grass-runner tiger beetle has already winked out of existence here). As far as I can tell, these two species are present on only a few small sand dunes, and these dunes are rapidly being overgrown by vegetation. It is likely various forms of erosion control will prevent the formation of new habitat in the near future, and thus it seems probable that we will lose these species from Alberta sometime in the next few decades.

The Coral Pink Sand Dunes in Utah, where tiger beetles and dune buggies are in constant legal conflict.

Admittedly, they are both at the limits of their species' geographic ranges, where they are naturally going to be rare. Populations of this sort sometimes come and go, and the loss of a species from Alberta does not have to be permanent. The painted turtle is another good example from our fauna, and I have heard that it seems to disappear from the Milk River drainage from time to time, only to reappear a few years later. Of course the Milk River is still there when they return, while the sand-dune habitats of the tiger beetles might not be so permanent.

It is also possible that these two tiger beetles might be living undetected on other dunes, or in small, unexplored sandy places away from main roads. I doubt it, however. I think we know enough about these beetles to be confidently skeptical. Anything is possible, mind you, and discoveries of new populations are a big part of what keeps tiger-beetlers going.

Entomologists involved with insect conservation have argued that tiger beetles, along with butterflies and dragonflies, are vastly over-represented in the lists of endangered insects. A rational approach to invertebrate conservation should, they say, give much more weight to thousands of species of small brown beetles, nondescript tiny flies and wasps, nematode worms, soil mites, and springtails.

From a scientific standpoint, this makes perfect sense, as does the notion that one should protect and preserve entire ecosystems, rather than becoming bogged down in the process of looking out for individual species on a planet where the number of species may lie somewhere in the tens of millions.

However, there is a problem with the rational approach. Simply put, very few people care about being rational. I think of it this way: insect conservation is a human activity, motivated by human concerns. We protect tiger beetles because we like them, not because they are demonstrably important to the rest of the world. We see their beauty, we know a great deal about their past and present distributions, and we choose them first when asked which insects most need protection from the actions of people. As well, conservation is based on the notion—often unstated—that we understand the world well enough to look after it. Protecting known, familiar species gives us the feeling that we know what we are doing.

I support this approach, irrational as it may be, since I like tiger beetles, but I do offer one note of caution. All too often, when something in nature comes under government protection, new regulations have the effect of "protecting" it from the very people who want most to help and study it. I would hate someday to find myself being warned not to catch, collect, photograph, approach, or otherwise harass the beetles, as if their rarity had also made them unusually sensitive to activities that don't seem to have harmed other tiger beetles elsewhere, at any point in recorded history.

If we choose to protect tiger beetles, let's protect them and enjoy them too. Let's think of them as a renewable resource, and a watchable wildlife resource. And let's not ignore the possibility that they might also be a sustainable resource that would allow limited, regulated collecting, given that they are more prolific than most of our sport fishes, game birds, and mammals. In today's North American society, we may not like the idea of an entomologist with a killing jar, but we should be open to sensible arguments and rely on evidence rather than gut feelings when restricting the activities of others.

Having said that, I don't want to end this book with a defense of insect collecting. Instead, I want to come back to the point on which we opened. There is a place for insect collecting, but its ability to inspire large numbers of people to a deep appreciation of nature has proven itself to be limited. It is therefore time to broaden our approach. In the past, amateur entomologists often thought of themselves as museum curators and taxonomists. Now, they seem happier just seeing, watching, and identifying things, the way birders do. In other words, it is time for us to try insect-watching, and what better place could there be to start than by admiring a few good tiger beetles?

Most Alberta sand dunes are being overgrown by plants, despite so-called droughts in recent years.

Many general naturalists, Chris Fisher among them, are taking up an interest in tiger beetles.

Afterword

AS THIS BOOK WAS GOING TO PRESS in the spring of 2000, my tiger-beetling friends and I noticed some very interesting changes in the *Cicindela* fauna of southern Alberta. First, there were not just a few, but dozens of blowout tigers on the sandy roads near Chappice Lake, north of Medicine Hat. I have seen old records of this species at this location, but I'll admit I was skeptical of them, since I've spent so much time here over the years without seeing a single representative of this species.

Then, at the Hilda Sand Dunes, the festive tiger beetle appeared, for the first time to my knowledge, in unbelievable numbers. Chris Schmidt called to tell me that he had also found a single individual of this species on the Gem Sand Dunes, confirming another old record that had puzzled me for years. Finally, on the dunes in Sceptre, Saskatchewan, where none of us had ever seen a beautiful tiger beetle, David Lawrie caught a lovely fresh individual, of the *gibsoni* race, typical of the Great Sand Hills to the south.

The past two winters have been unusually mild, and in the absence of other evidence, or other obvious hypotheses, I'd like to suggest that this is probably the factor that has allowed these beetles to expand their ranges into what are usually marginal or unsuitable habitats. The message I take from these records is simple: nature in general, and *Cicindela* in particular, represents an ever-changing, swirling pattern, sometimes cyclical but never static. The joy of writing a book such as this one, and "documenting" one aspect of the fauna of Alberta, is that things will surely change with time (even short periods of time), and the next person who comes along will have a whole new task before him or her, with this book as a starting point—a historical record of a period that was indeed well studied, but a mere instant in the ever-changing natural "history" of our province.

The ghost tiger beetle is one of the most recent additions to the provincial checklist.

APPENDIX 1

Checklist of Alberta Tiger Beetles

TIGER BEETLES, family Cicindelidae
Familiar tiger beetles, genus *Cicindela*

TINY TIGER BEETLES, subgenus *Cylindera*
grass-runner tiger beetle, *Cicindela cinctipennis*
  cinctipennis race
  imperfecta race

CURLICUE TIGER BEETLES, subgenus *Ellipsoptera*
salt creek tiger beetle, *Cicindela nevadica*
  knausi race, *C. n. knausi*
ghost tiger beetle, *Cicindela lepida*
  brown morph
  green morph

AMERICAN TIGER BEETLES, subgenus *Cicindelidia*
backroad tiger beetle, *Cicindela punctulata*
  punctulata race, *C. p. punctulata*

TEMPERATE TIGER BEETLES, subgenus *Cicindela*

A. THE SYLVAN GROUP (*C. sylvatica* species group)
  long-lipped tiger beetle *Cicindela longilabris*
    longilabris race
    racial intergrades (*C. l. longilabris X C. l. laurentii X C. l. perviridis*)
      black morph
      olive morph
      green morph
  black-bellied tiger beetle, *Cicindela nebraskana*

B. THE MARITIME GROUP (*C. maritima* species group)
  bronzed tiger beetle, *Cicindela repanda*
    repanda race
  twelve-spot tiger beetle, *Cicindela duodecimguttata*
  Pacific tiger beetle, *Cicindela oregona*
    racial intergrades (*C. o. oregona X C. o. guttifera*)
  Pacific/twelve-spot hybrids (*C. oregona X C. duodecimguttata*)
  beach tiger beetle, *Cicindela hirticollis*
    shelfordi race
    athabascensis race
      brown morph
      blue morph
      green morph
    populations of uncertain status

sandy tiger beetle, *Cicindela limbata*
    nympha race
        brown morph
        green morph
    hyperborea race
        brown morph
        green morph

THE COWPATH GROUP (*C. purpurea* species group)
claybank tiger beetle, *Cicindela limbalis*
    limbalis race
        bright morph
        dull morph
cowpath tiger beetle, *Cicindela purpurea*
    auduboni race
        black morph
        green morph
badlands tiger beetle, *Cicindela decemnotata*

THE "UNGROUPED" SPECIES
beautiful tiger beetle, *Cicindela formosa*
    populations of uncertain status
festive tiger beetle, *Cicindela scutellaris*
    scutellaris race
        bright morph
        dull morph
shiny tiger beetle, *Cicindela fulgida*
    greenish red morph
    reddish green morph
    blue morph
    green morph
    purple morph
blowout tiger beetle, *Cicindela lengi*
    versuta race
        red morph
        blue morph
        green morph
        black morph
oblique tiger beetle, *Cicindela tranquebarica*
    kirbyi race
        black morph
        olive morph
        green morph
    borealis race
        black morph
        olive morph
        green morph

APPENDIX 2

Key to the Adult Tiger Beetles of Alberta

The traditional way to identify beetles is to "key them out." A key is simply a string of choices that lead you through a proccess of elimination until only one species remains. To some people, a key is a sign of intellectual rigour, while to others it represents the constraining process of "linear thinking" that makes science so apparently unfeeling. Bird watchers do not use keys, so they are not a neccessity for animal identification. For those who like them, however, I have written one. It is not entirely traditional in form, since at some points there are three rather than two choices. I also use words to explain what to do next, rather than just numbers. You should have no trouble finding pictures in the book to illustrate the features mentioned in the key. This will come naturally if you check each identification against the appropriate species treatment. As usual, start at 1, and follow on from there. It may help to remember that there is no information in the key that does not appear elsewhere in the text.

1

Are the wing covers mostly white, and are the white markings somewhat run together? If so, go to 2.

Or do they have no white markings at all? If so, go to 3.

Or are they mostly a darker colour, with some white markings? If so, go to 4.

2 (from 1)

Are the legs dark brown and green in colour? Then it is the ***nympha*** **race of the sandy tiger beetle, *Cicindela limbata nympha.***

Or are they tan in colour? Then it is the **ghost tiger beetle, *Cicindela lepida.***

3 (from 1)

Are the wing covers entirely red and green? Then it is the **festive tiger beetle, *Cicindela scutellaris.***

Or is the beetle black, with a black underside? Then it is the **black-bellied tiger beetle, *Cicindela nebraskana.***

Or is the beetle black, olive, or green with an iridescent underside that is blue, purple, or green? Then it is the **long-lipped tiger beetle, *Cicindela longilabris.***

4 (from 1)

Is the beetle small (10 mm long or less)? Then it is the **grass-runner tiger beetle, *Cicindela cinctipennis.***

Or is it larger? If so, go to 5.

5 (from 4)

Is the shoulder mark coat-hook shaped or a straight bar, perpendicular to the line where the wing covers meet? If so, go to 6.

Or is it more or less finger-shaped, and obliquely directed toward the rear of the beetle? If so, go to 7.

Or is it a crescent, a dot, a teardrop, or not present at all? If so, go to 10.

6 (from 5)

Are the body hairs flattened against the body? Then it is the **salt creek tiger beetle, *Cicindela nevadica*.**

Or do the hairs stand out from the body, especially on the forehead? Then it is the **beach tiger beetle, *Cicindela hirticollis*.**

7 (from 5)

Is the beetle shinier than most other tiger beetles, and does the tip of the shoulder mark touch or nearly touch the middle band? If so, it is the **shiny tiger beetle, *Cicindela fulgida*.**

Or is it of average shininess, regardless of the shape of the shoulder mark? If so, go to 8.

8 (from 7)

Is the shoulder mark narrowed to the rear of the shoulder angle? If so, go to 9.

Or is the shoulder mark not noticeably narrowed to the rear of the shoulder angle? Then it is the **blowout tiger beetle, *Cicindela lengi*.**

9 (from 8)

Is the beetle small (10–12 mm long)? If so, it is the ***hyperborea* race of the sandy tiger beetle, *Cicindela limbata hyperborea*.**

Or, is it larger (13–15 mm long)? Then it is the **oblique tiger beetle, *Cicindela tranquebarica*.**

10 (from 5)

Is the beetle large (15-17 mm) and red, with a blue underside? Then it is the **beautiful tiger beetle, *Cicindela formosa*.**

Or, is it smaller, and differently coloured? If so, go to 11.

11 (from 10)

Is the upper lip about half as long as it is wide? If so, go to 12.

Is the upper lip less than half as long as it is wide? If so, go to 13.

12 (from 11)

Is the beetle mainly black, with a black underside? Then it is the **black-bellied tiger beetle, *Cicindela nebraskana*.**

Or is it mainly black, olive, or green with an iridescent underside that is blue, purple, or green? Then it is the **long-lipped tiger beetle, *Cicindela longilabris*.**

13 (from 11)

Is the shoulder mark a crescent, extending from the shoulder itself down along the margin of the wing cover and then inward? If so, go to 14.

Or does the shoulder mark consist only of a dot below the shoulder, with or without a white mark on the shoulder itself? If so, go to 15.

Or is the shoulder mark absent altogether, or reduced to a dot at the shoulder angle? If so, go to 19.

14 (from 13)

Does the shoulder mark connect with, or come close to connecting with, the middle band along the edge of the wing cover? Then it is the **bronzed tiger beetle, *Cicindela repanda*.**

Or is the shoulder mark clearly separated from the middle band along the edge of the wing cover? Then it is the **twelve-spot tiger beetle, *Cicindela duodecimguttata*.**

15 (from 13)

Is there a row of tiny green punctures running the length of each wing cover, set against a black background? Then it is the **backroad tiger beetle, *Cicindela punctulata*.**

Or is there no such row of punctures? If so, go to 16.

16 (from 15)

Is the forehead quite hairy? If so, go to 17.

Or are there only a few hairs on the forehead, just over each eye? Then it is the **Pacific tiger beetle, *Cicindela oregona*.**

17 (from 16)

Is the beetle mostly brown to the unaided eye? Then it is the **twelve-spot tiger beetle, *Cicindela duodecimguttata*.**

Or is it green, greenish, or red? If so, go to 18.

18 (from 17)

Are the head and prothorax all greenish or blue-greenish? Then it is the **badlands tiger beetle, *Cicindela decemnotata*.**

Or are the head and prothorax at least partly red with blue-green trim? Then it is the **claybank tiger beetle, *Cicindela limbalis*.**

19 (from 13)

Is there a row of tiny green punctures running the length of each wing cover, set against a black background? Then it is the **backroad tiger beetle, *Cicindela punctulata*.**

Or, is there no such row of punctures? If so, go to 20.

20 (from 19)

Is the beetle brown or blue-green, with a contiguous apical mark and a well-formed, curvaceous middle band? Then it is the ***athabascensis* race of the beach tiger beetle, *Cicindela hirticollis*.**

Or is it black or green, with the apical mark consisting of a white dot as well as a white tip to the wing cover, and with a short, simple middle band? Then it is the **cowpath tiger beetle, *Cicindela purpurea*.**

To really enjoy tiger-beetling, you need a good net and sturdy outdoor footwear.

APPENDIX 3

Helpful Sources for Tiger-Beetlers

SUPPLIES

ATELIER JEAN PAQUET: 3 rue du Côteau, P.O. Box 953, Pont-Rouge, Québec, Canada G3H 2E1. E-mail jeanpaquet@webnet.qc.ca. Insect nets and collecting supplies.

BIO QUIP INC.: 17803 LaSalle Avenue, Gardena, California, USA, 90248-3602. phone (310) 324 0620, fax (310) 324 7931. e-mail: bioquip @aol.com. Good selection of nets, other entomological equipment, books, and videos.

YOUNG ENTOMOLOGISTS SOCIETY: see below

JOURNALS, WEBSITES AND SOCIETIES

CICINDELA. This is the only journal entirely devoted to the subject of tiger beetles. Subscription is US $7.00 per year, payable to Ronald L. Huber, Senior Editor, 4637 W. 69th terrace, Prairie Village, Kansas, USA, 66208.

THE COLEOPTERISTS SOCIETY. An international society devoted to the study of beetles, that publishes "The Coleopterists Bulletin." Membership is US $30.00 year, payable to the society's treaurer, currently Terry Seeno, CDFA-PPD, 3294 Meadowview Road, Sacramento, California, 95832-1448, U.S.A.. e-mail: tseeno@ns.net.

YOUNG ENTOMOLOGISTS SOCIETY: an international group for amateur entomologists, that publishes a variety of useful and entertaining books and newletters, as well as operating the Minibeast Merchandise Mall, an internet entomology store. Membership dues are reasonable, but depend on your age, profession, and whether you live in Canada or the USA.
Address: 1915 Peggy Place, Lansing, Michigan, USA, 48910-2553.
website: http://members.aol.com/YESbugs/bugclub.html
e-mail: YESbugs@aol.com

ENTOMOLOGICAL SOCIETY OF ALBERTA: a group of entomologists (primarily professional) that meets once a year for three days, and publishes abstracts from the papers presented at the meetings. Amateurs are welcome. Membership is $10 per year ($5 for students), payable to the Secretary of the Entomological Society of Alberta, c/o Department of Biological Sciences, University of Alberta, Edmonton, Alberta, T6G 2E9.

ENTOMOLOGICAL SOCIETY OF CANADA: a group of entomologists that meets once a year, and publishes a newsletter as well as "The Canadian Entomologist," and "Memoirs of the Canadian Entomological Society." Amateurs are welcome. Membership fee depends on whether you live in Canada or not, are a student or not, and which publications you want to receive.

Address: 1320 Carling Avenue, Ottawa, Ontario, K1Z 7K9.
website: http://www.biology.ualberta.ca/esc.hp/homepage.htm

THE E.H. STRICKLAND ENTOMOLOGICAL MUSEUM houses a research collection of tiger beetles and other insects and has a great website.
Address: Room 218, Earth and Atmosphereic Sciences Building, University of Alberta, Edmonton, Alberta T6G 2E1.
website: http://www.biology.ualberta.ca/uasm.html

Glossary

ABDOMEN The hindmost division of the body, behind the last leg-bearing segment of the thorax.

ADEPHAGA The carnivorous suborder of beetles.

ADEPHAGAN A member of the Adephaga.

APICAL MARK The white markings nearest the tip of the wing cover.

CARABIDAE The beetle family that includes ground beetles, and according to some, the tiger beetles as well.

CICINDELOPHILE Latinized form of "tiger-beetler."

COLEOPTERA The insect order that includes beetles

COLEOPTERIST Someone who studies beetles.

COUPLING SULCUS A groove in the side of the female thorax, in which the male's mandibles fit snugly while mating.

ELYTRA Plural of elytron.

ELYTRON A wing cover, which is a modified front wing.

EVOLUTION The process of organic descent with modification, usually over very long periods of time.

FAMILY A group of related genera and the species they contain.

GENERA Plural of genus.

GENUS A group of related species.

HUMERAL LUNULE Shoulder mark.

HYBRID The resulting offspring from the interbreeding of two or more species.

INSECT The class of animals that possess an external skelton, three pairs of jointed legs, and three main body divisions: the head, thorax, and abdomen.

INSTAR The three sub-divisions of the larval stage, separated by sheddings of the larval skin.

INTERGRADES Populations formed by the interbreeding of two or more subspecies.

LARVA The immature stage of the life cycle after the egg and before the pupa.

MANDIBLES The paired jaws, along with all of their tooth-like projections.

MIDDLE BAND The light marking that extends into the middle of the wing cover from the outer margin, bending twice and taking the shape of a flexed human leg in profile.

NATURAL SELECTION The process whereby traits that cause their possessors to leave proportionately more offspring than others, tend to increase in frequency in a population.

ORDER A group of related families, and the species they contain.

PRONOTUM The top of the prothorax.

PROTHORAX The body segment behind the head, which bears the front legs.

SETAE Insect hairs that are set in sockets.

SPECIES GROUP A group of related species within a subgenus.

SPECIES A group of organisms that interbreed in nature, look more-or-less alike, and share a common evolutionary heritage.

SUBFAMILY A group of related genera (and the species they contain) within a family.

SUBGENUS A group of related species within a genus.

SUBSPECIES A geographic race within a species, usually with distinctive markings.

TARSI Plural of tarsus.

TARSOMERE An individual toe segment.

TARSUS The last five "toe" segments of the leg, beyond the tibia, taken together as a group

TIGER-BEETLER Anyone with a passion for tiger beetles.

TRIBE A group of related genera (and the species they contain) within a subfamily.

References

Acorn, John H. 1991. Habitat associations, adult life histories, and species interactions among sand dune tiger beetles in the southern Canadian prairies (Coleoptera: Cicindelidae). Cicindela. 23(2/3): 17-48.

Acorn, John H. 1992. The Historical Development of Geographic Color Variation among Dune *Cicindela* in Western Canada (Coleoptera: Cicindelidae). In Noonan, G.R., G.E. Ball, and N.E. Stork, eds. The Biogeography of Ground Beetles of Mountains and Islands. Intercept Ltd., Hampshire. pp. 217-233.

Acorn, John H. 1988. Mimetic tiger beetles and the puzzle of cicindelid coloration. Coleopterists Bulletin. 42: 29-33.

Acorn, John H. 1994. Tiger Beetles (Coleoptera: Cicindelidae) of the Lake Athabasca sand dunes—an intriguing northern assemblage. Cicindela. 26: 9-16.

Criddle, Norman. 1907. Habits of some Manitoba tiger beetles (Cicindelidae). The Canadian Entomologist. 39: 105-114.

Criddle, Norman. 1910. Habits of some Manitoba tiger beetles, No. 2. (Cicindelidae). The Canadian Entomologist. 42: 9-15.

Freitag, Richard. 1965. A revision of the North American species of the *Cicindela maritima* group with a study of hybridization between *Cicindela duodecimguttata* and *oregona*. Quaestiones Entomologicae. 1: 87-170.

Gaumer, Grant C. 1977. The variation and taxonomy of *Cicindela formosa* Say (Coleoptera: Cicindelidae). Ph.D. Dissertation. Texas A & M University, College Station, Texas. 253 pp.

Hilchie, Gerald J. 1985. The tiger beetles of Alberta (Coleoptera: Carabidae, Cicindelini). Quaestiones Entomologicae. 21: 319-347.

Kniseley, C. Barry, and Tom D. Schultz. 1997. The Biology of Tiger Beetles and a Guide to the Species of the South Atlantic States. Virginia Museum of Natural History. Special Publication Number 5. 210 pp.

Leonard, Jonathan G. and Ross T. Bell. 1999. Northeastern Tiger Beetles: A Field Guide to Tiger Beetles of New England and Eastern Canada. CRC Press. 176 pp.

Morgan, K.R. 1985. Body temperature regulation and terrestrial activity in the ectothermic tiger beetle *Cicindela tranquebarica*. Physiological Zoology 58: 29-37.

Pearson, David L. 1988. Biology of tiger beetles. Annual Review of Entomology. 33: 123-147.

Shincariol, Larry A., and Richard Freitag. 1991. Biological character analysis, classification, and history of the North American *Cicindela splendida* Hentz group taxa (Coleoptera: Cicindlidae). The Canadian Entomologist 123: 1327-1353.

Spanton, Timothy G. 1988. The *Cicindela sylvatica* group: geographical variation and classification of the Nearctic taxa, and reconstructed phylogeny and geographical history of the species (Coleoptera: Cicindelidae). Quaestiones Entomologicae. 24: 51-161.

Wallis, J.B. 1961. The Cicindelidae of Canada. University of Toronto Press. 72 pp.

Werner, Karl. 1994. The Beetles of the World, Volume 18: Cicindelidae 3. Sciences Nat, France. 196 pp.

Werner, Karl. 1994. The Beetles of the World, Volume 20: Cicindelidae 4. Sciences Nat, France. 196 pp.

Wiesner, Jürgen. 1992. Verzeichnis der Sandlaufkäfer der Welt. Checklist of the Tiger Beetles of the World. Erna Bauer, Keltern-Weiler, Germany. 346 pp.

Willis, Harold L. 1967. Bionomics and zoogeography of tiger beetles of saline habitats in the central United States (Coleoptera: Cicindelidae). University of Kansas Science Bulletin. 48: 145-313.

A Gallery of Tiger Beetles

Scale: 1.3 times lifesize

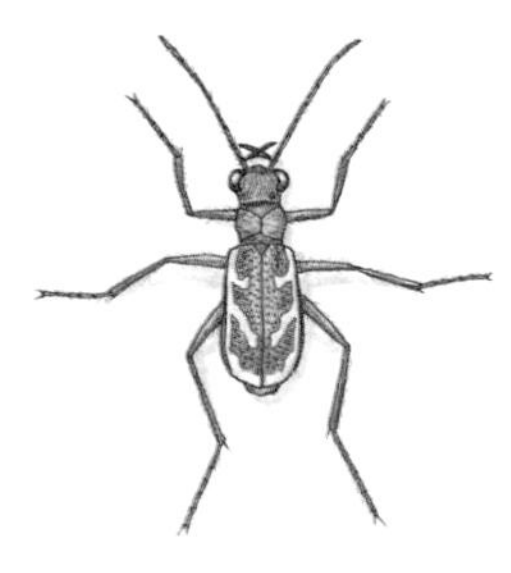

GRASS-RUNNER TIGER BEETLE
(*Cicindela cinctipennis cinctipennis*)
(p. 20)

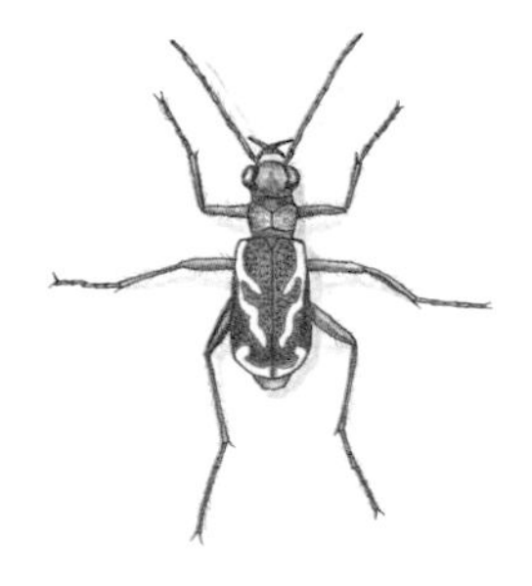

GRASS-RUNNER TIGER BEETLE
(*Cicindela cinctipennis imperfecta*)
(p. 20)

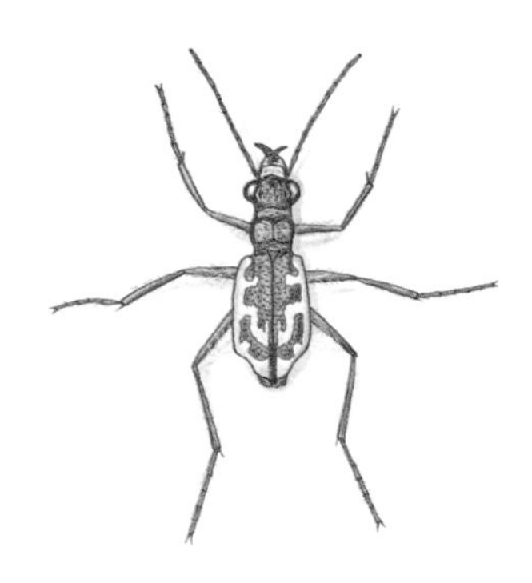

SALT CREEK TIGER BEETLE
(*Cicindela nevadica knausi*) (p. 26)

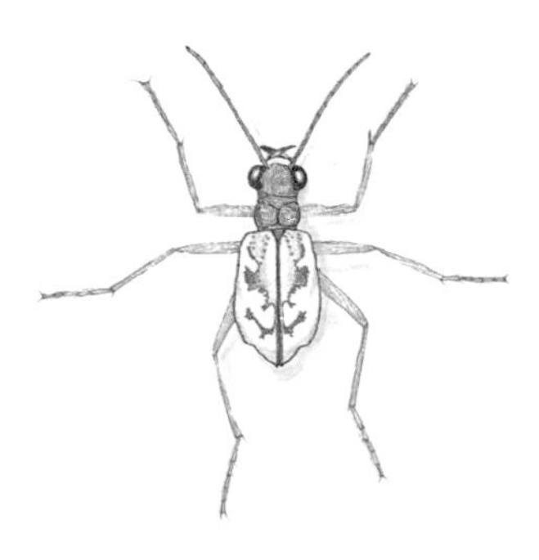

GHOST TIGER BEETLE
(*Cicindela lepida*) (p. 28)

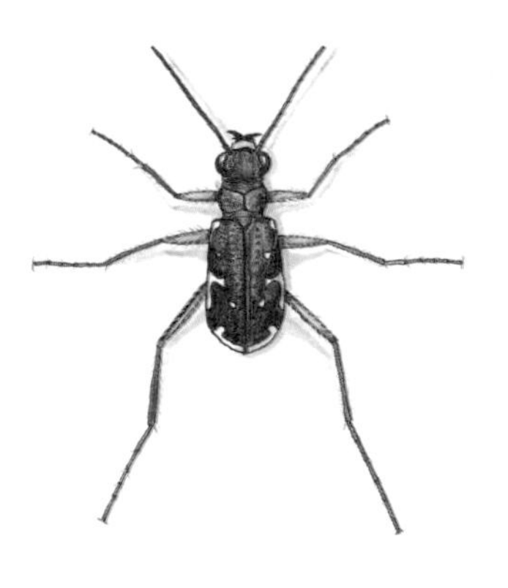

BACKROAD TIGER BEETLE
(*Cicindela punctulata*) (p. 32)

BLACK-BELLIED TIGER BEETLE
(*Cicindela nebraskana*) (p. 37)

LONG-LIPPED TIGER BEETLE
(*Cicindela longilabris*) (p. 39)

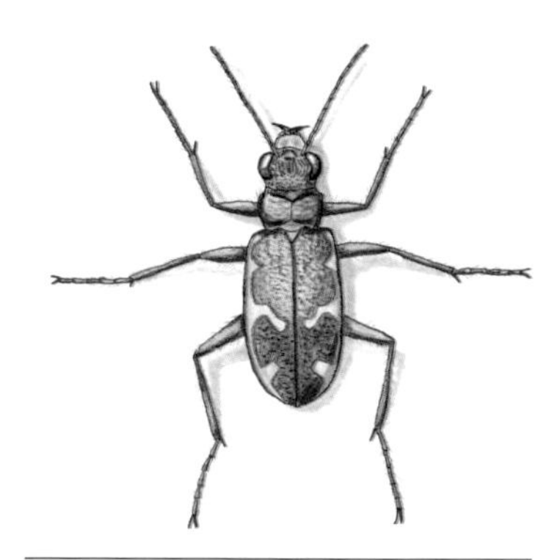

LONG-LIPPED TIGER BEETLE
(*Cicindela longilabris*)
intergrade green morph (p. 39)

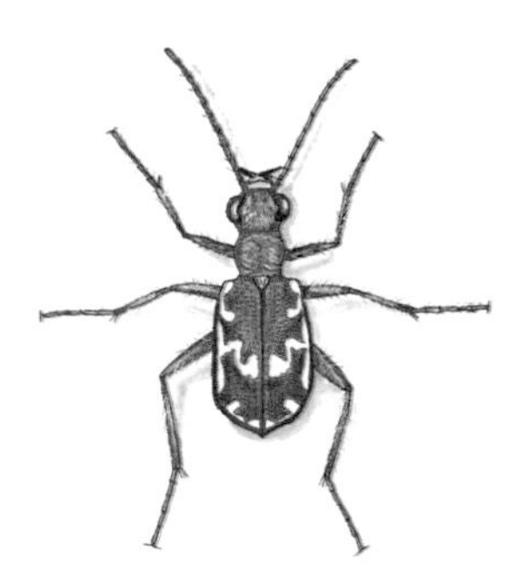

BRONZED TIGER BEETLE
(*Cicindela repanda*) (p. 43)

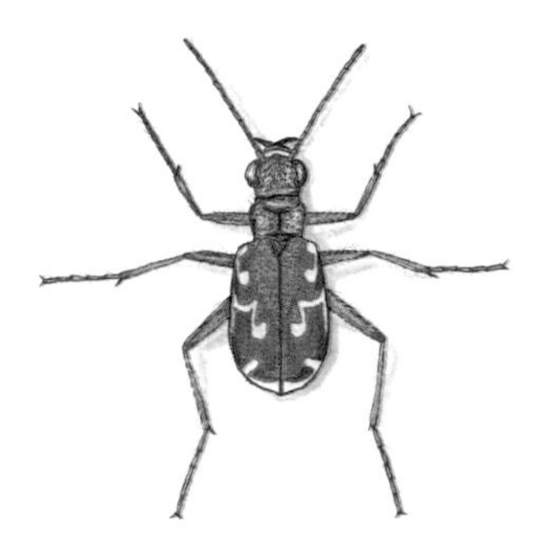

TWELVE-SPOT TIGER BEETLE
(*Cicindela duodecimguttata*) (p. 45)

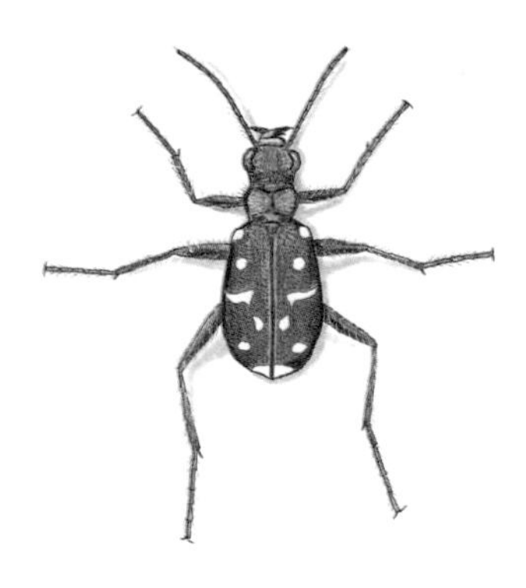

PACIFIC TIGER BEETLE
(*Cicindela oregona*) (p. 47)

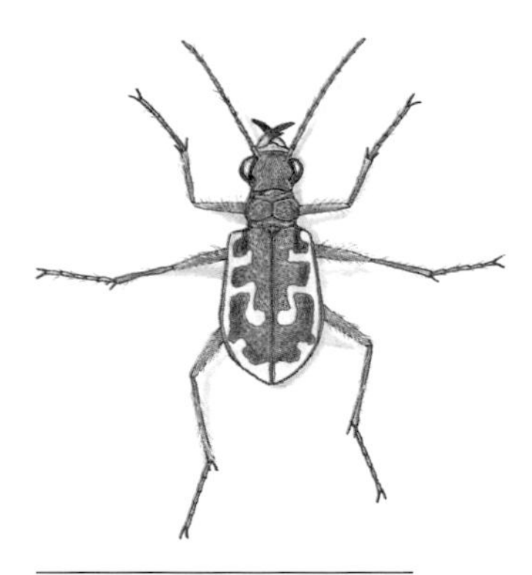

BEACH TIGER BEETLE
(*Cicindela hirticollis*) (p. 49)

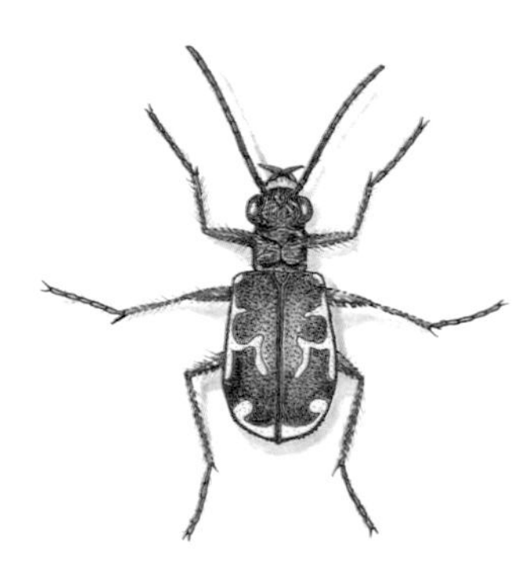

BEACH TIGER BEETLE
(*Cicindela hirticollis athabascensis*)
brown morph (p. 49)

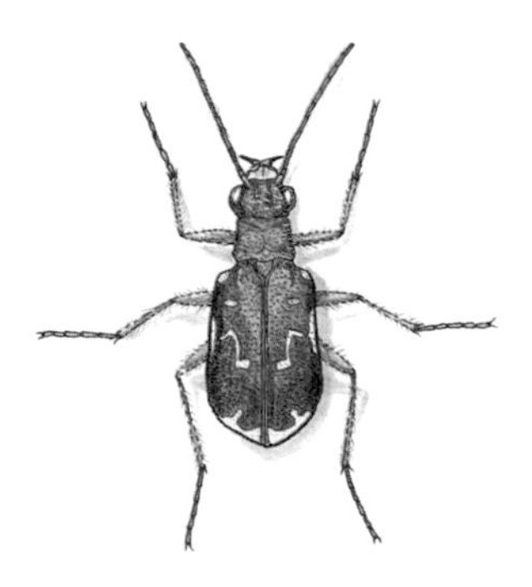

BEACH TIGER BEETLE
(*Cicindela hirticollis athabascensis*)
blue form (p. 49)

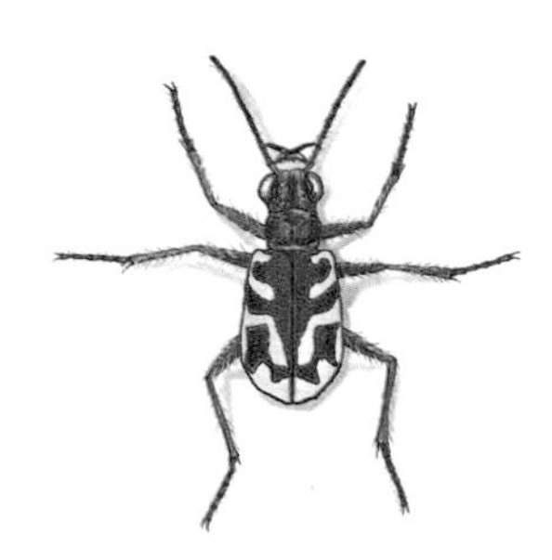

SANDY TIGER BEETLE
(*Cicindela limbata hyperborea*) (p. 52)

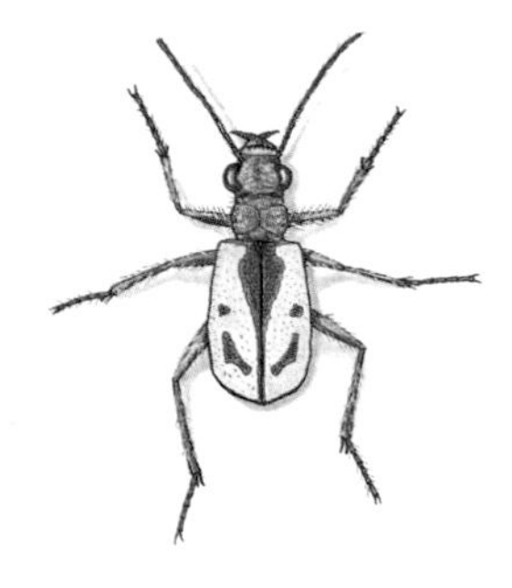

SANDY TIGER BEETLE
(*Cicindela limbata nympha*) (p. 52)

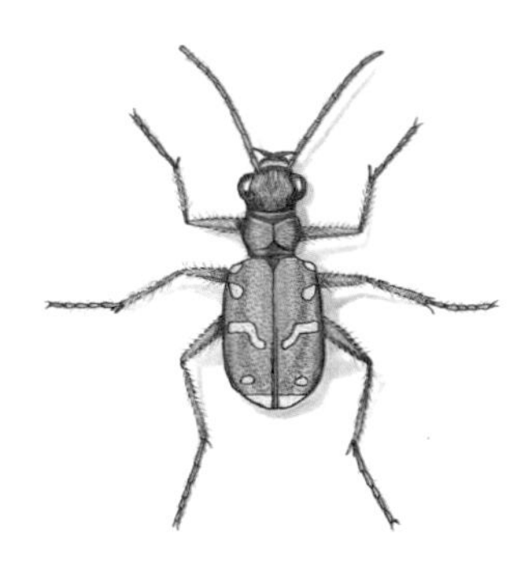

CLAYBANK TIGER BEETLE
(*Cicindela limbalis*)
green morph (p. 56)

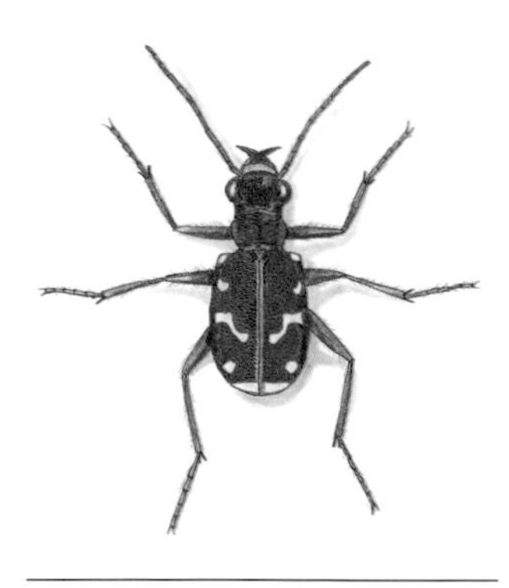

CLAYBANK TIGER BEETLE
(*Cicindela limbalis*)
red morph (p. 56)

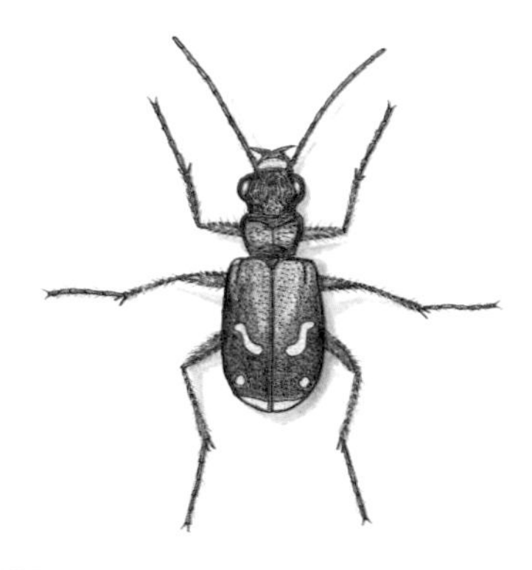

COWPATH TIGER BEETLE
(*Cicindela purpurea auduboni*)
black morph (p. 59)

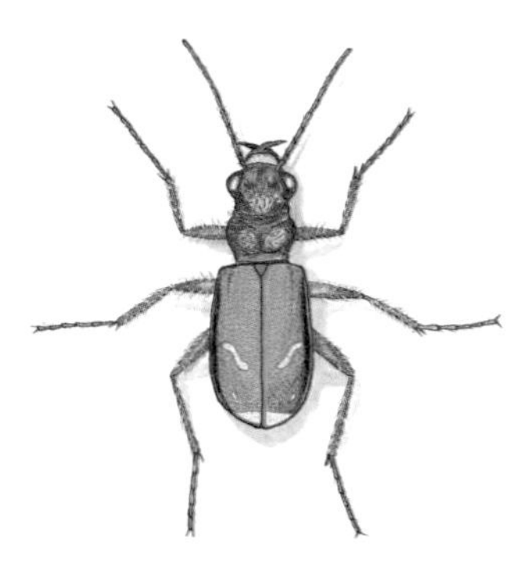

COWPATH TIGER BEETLE
(*Cicindela purpurea auduboni*)
green morph (p. 59)

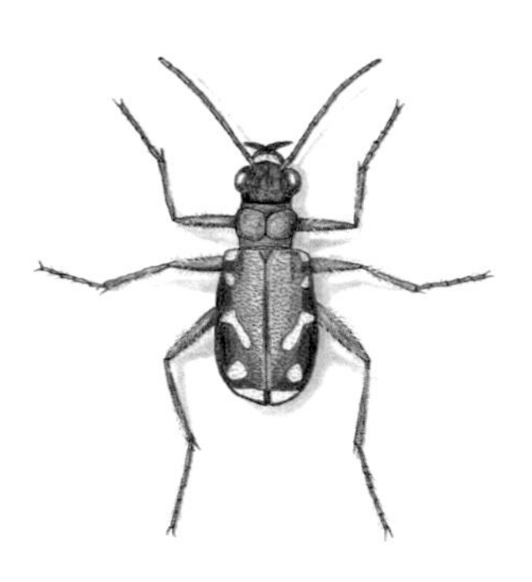

BADLANDS TIGER BEETLE
(*Cicindela decemnotata*) (p. 62)

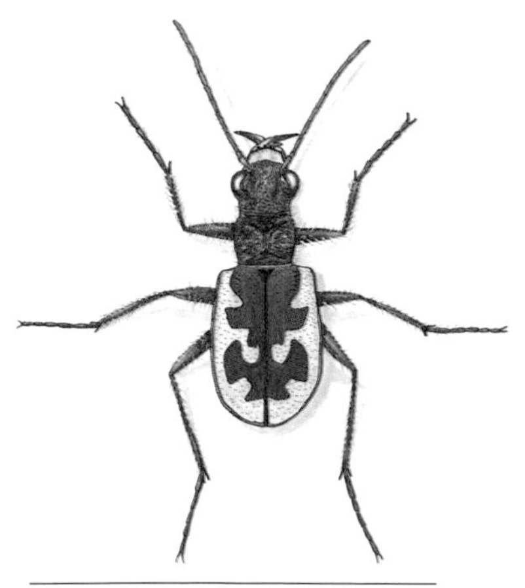

BEAUTIFUL TIGER BEETLE
(*Cicindela formosa*) (p. 66)

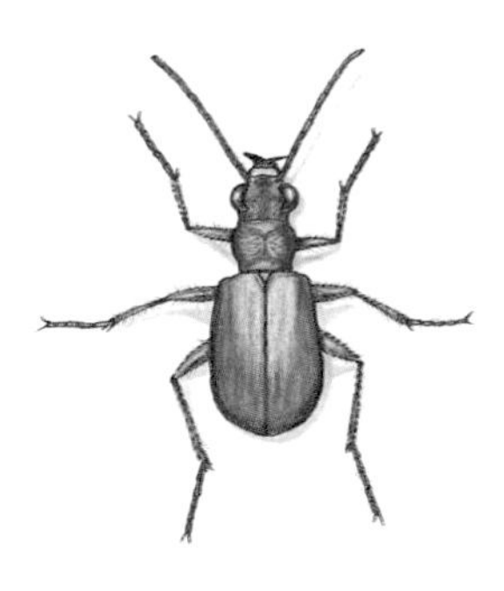

FESTIVE TIGER BEETLE
(*Cicindela scutellaris*) (p. 69)

SHINY TIGER BEETLE
(*Cicindela fulgida "westbournei"*)
blue morph (p. 72)

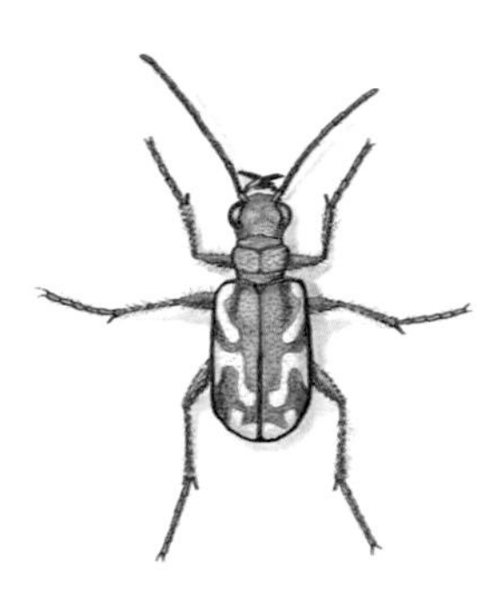

SHINY TIGER BEETLE
(*Cicindela fulgida "westbournei"*)
green morph (p. 72)

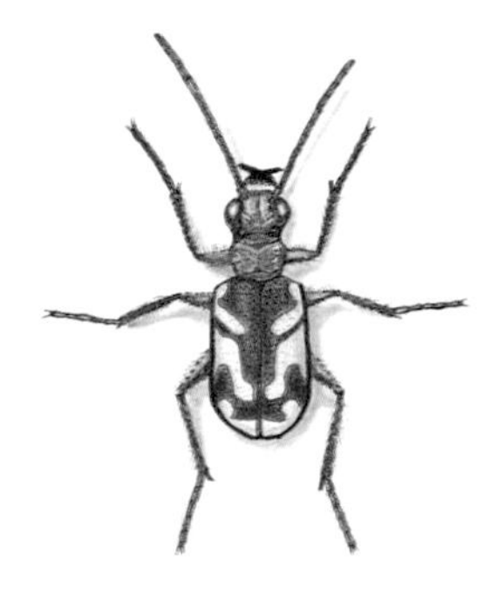

SHINY TIGER BEETLE
(*Cicindela fulgida "westbournei"*)
greenish red morph (p. 72)

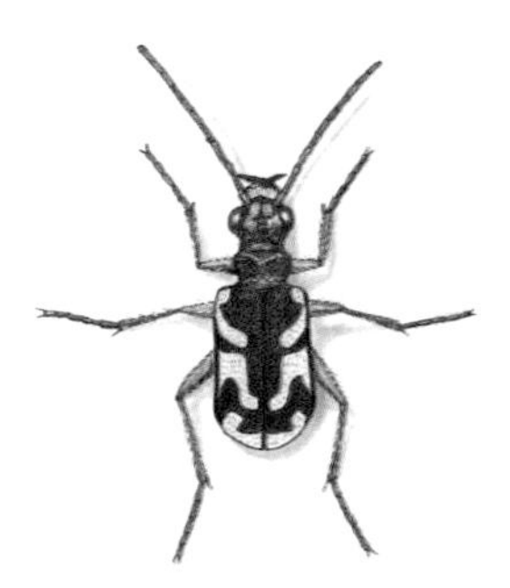

SHINY TIGER BEETLE
(*Cicindela fulgida "westbournei"*)
purple morph (p. 72)

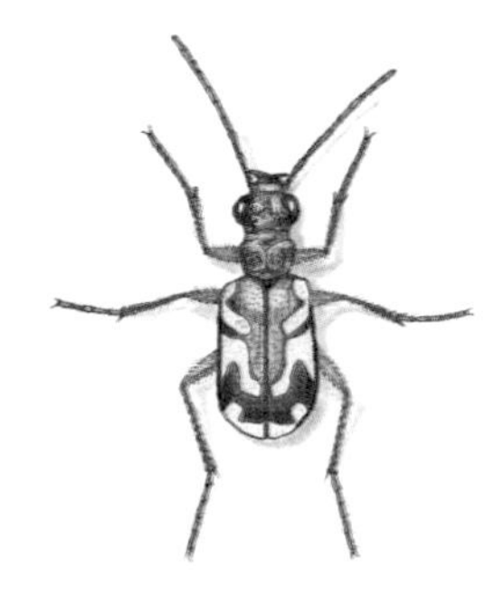

SHINY TIGER BEETLE
(*Cicindela fulgida "westbournei"*)
reddish green morph (p. 72)

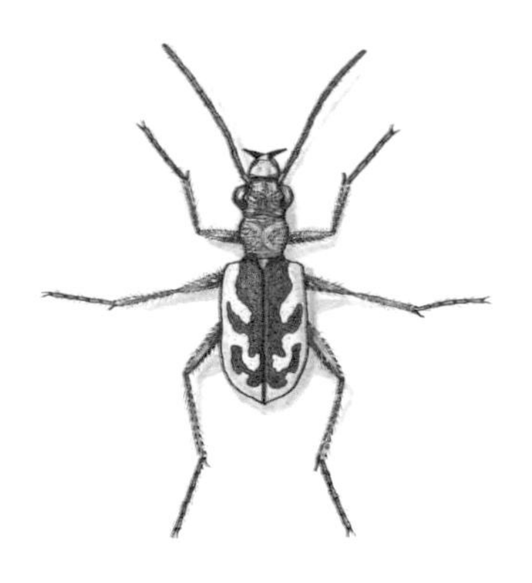

BLOWOUT TIGER BEETLE
(*Cicindela lengi versuta*)
red (common) morph (p. 75)

BLOWOUT TIGER BEETLE
(*Cicindela lengi versuta*)
dark morph (p. 75)

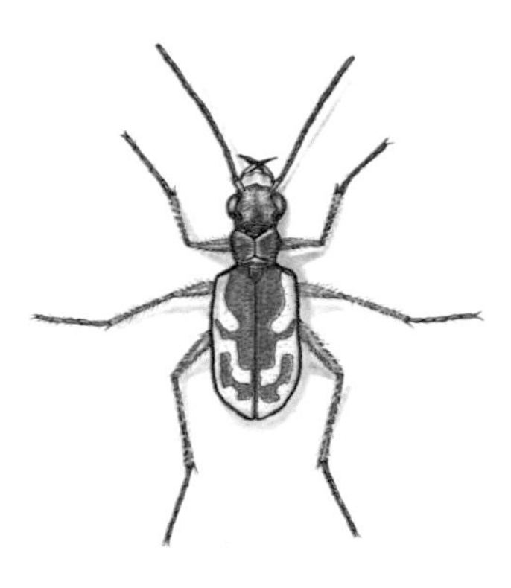

BLOWOUT TIGER BEETLE
(*Cicindela lengi versuta*)
blue morph (p. 75)

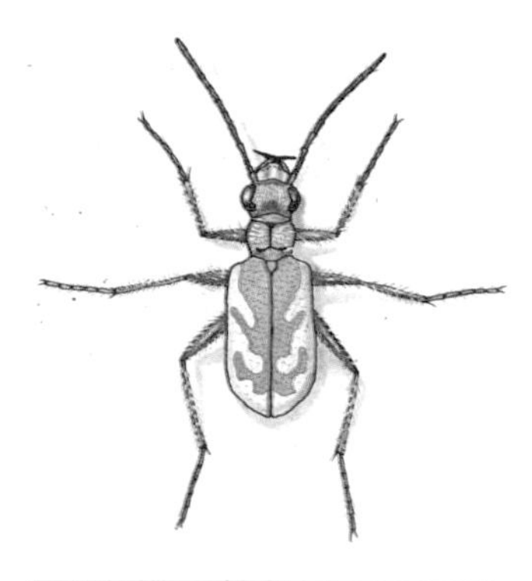

BLOWOUT TIGER BEETLE
(*Cicindela lengi versuta*)
green morph (p. 75)

OBLIQUE TIGER BEETLE
(*Cicindela tranquebarica borealis*)
(p. 79)

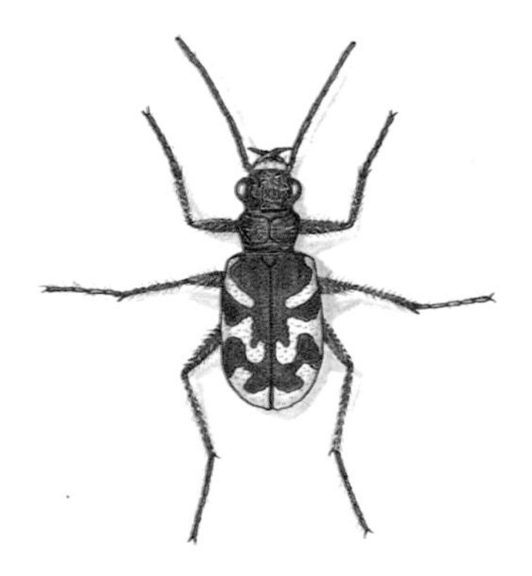

OBLIQUE TIGER BEETLE
(*Cicindela tranquebarica kirbyi*)
(p. 79)